Date label / bar-code

A. I. Beltzer

Acoustics of Solids

With 90 Figures

Springer-Verlag
Berlin Heidelberg New York
London Paris Tokyo

Professor Abraham I. Beltzer

Holon Institute for Technological Education
Affiliated with Tel Aviv University
The Holon Campus
52 Golomb Street
Holon 58102
Israel

ISBN 3-540-18888-6 Springer-Verlag Berlin Heidelberg NewYork
ISBN 0-387-18888-6 Springer-Verlag NewYork Heidelberg Berlin

Library of Congress Cataloging-in-Publication Data
Beltzer, Abraham I.: Acoustics of solids.
Bibliography: p.
Includes indexes.
1. Acoustic surface waves
2. Solids Acoustic properties.
3. Wave guides.
I. Title.
QC176.8.A3B45 1988 530.4'1 88-4611
ISBN 0-387-18888-6 (U.S.)

Dataconversion: EDP-Consulting Mattes, Heidelberg
Offsetprinting: Color-Druck, G. Baucke, Berlin; Bookbinding: Lüderitz & Bauer, Berlin
2161/3020 5 4 3 2 1 0

To Dina, Ruth and Orna

Preface

Wave propagation in solids has for a long time been one of the favorite activities of applied mathematicians, but the field of applications has been quite narrow, consisting principally of seismology and geophysics. Technological developments in composite materials, non-destructive testing, and signal processing as well as biomedical applications, have stimulated wide-ranging engineering investigations of heterogeneous, anisotropic media and surface waves of different types, which have changed the situation. Wave propagation in solids is now of considerable importance in a variety of applications. The acoustics of solid media has thus become an engineering discipline like the acoustics of fluids.

This book does not purport to provide an exhaustive treatment of waves in solids, which would be an impossible task, and focuses mainly on the principle approaches and physical meanings. It presents many of the key results in this field and interprets them from a unified, engineering viewpoint. The conceptual importance and relevance for applications were therefore the prevailing criteria in selecting the topics.

This book deals almost exclusively with the linear theory. Included are body and surface waves in elastic, viscoelastic, and piezoelectric media and waveguides, with emphasis on the effects of inhomogeneity and anisotropy. The book differs in many aspects from the other monographs dealing with wave propagation in solids. It focuses on physically meaningful theoretical models, a broad spectrum of which is covered, and not on mathematical techniques. For these a reader should consult excellent books available. Some of the results are given for the first time in the monographical literature. Both, exact and approximate approaches, are discussed. While the subject is advanced, the presentation is at an intermediate level of mathematical complexity, which is perhaps more justified in mechanics than in other sciences. Nevertheless, a preliminary background in traditional engineering courses, like elasticity, vibrations, and materials, appears desirable for the understanding of the subject.

Chapter 1 begins with a general discussion of the dynamic response of materials and their structures. This is followed by a presentation of the conservation laws, basic mechanical models, and concepts of waves, so as to provide the ground work for the subsequent treatment.

Foundations of the theory of bulk elastic and viscoelastic waves in an unbounded isotropic medium are given in Chapter 2. The analysis of radiation from cavities, from rigid inclusions, and from imperfections is as important for appreciating the subject as the elementary oscillator for the theory of vibrations.

The basic results concerning waves in anisotropic media are given in Chapter 3, a substantial portion of which is devoted to piezoelectric waves. Simple cases of symmetry have been chosen so as to avoid tedious calculations. Nevertheless, they show quite clearly a variety of new emerging phenomena.

Chapter 4 deals with effects of boundaries and properties of waveguides, which are investigated through exact as well as approximate theories. As a convenient means of investigating a forced response the technique of modal superposition is given. For vibrations of various continuous systems and related topics the reader is referred to monographs devoted to this subject.

The assumption of homogeneity is removed in Chapter 5, giving rise to complicated new phenomena. Investigations of diffraction by an inclusion are followed by an analysis of wave propagation in periodic and random composites. Alternative approximate methods and their interpretations are given. This chapter contains also some of the recently obtained results concerning the effective response of stochastic multiphase media.

Finally, exercises and references extend the scope of the book, which should be useful in view of the huge literature available on the dynamic response of solids. References, which are by no means exhaustive, are collected at the end of each chapter and are accompanied by remarks aiming to facilitate a proper choice of further sources. Hopefully, this arrangement makes reading more comfortable. Unfortunately, it was not always possible to follow standard notations. Letters in boldface denote vectors or tensors and the usual summation convention with respect to repeated indices applies.

This monograph is an outgrowth of lectures taught at the Holon Institute for Technological Education, affiliated with Tel-Aviv University, and the Division of Solid Mechanics, Department of Mechanical Engineering, Lulea University of Technology. Parts of the book were written while I was on leave at the Universities of Oklahoma and Paderborn. The availability of rich literature concerning wave propagation and discussions with my colleagues Professors C.W. Bert, D.M. Egle, K. Herrmann, B. Lundberg and N. Rudy have made this undertaking particularly enjoyable. At the stage of final editing I was on leave at the Department of Mechanical Engineering, University of Alberta, Edmonton, and shared views with Professor J.B. Haddow. The collaboration of my students, their humor, and especially their patience are highly appreciated.

A.I. Beltzer
Edmonton, Alberta

Contents

Chapter I

Elements of Material Structure and Solid Dynamics

This chapter outlines the key concepts relevant to the subject, provides the relations necessary for the subsequent "build-up" of the theory, and introduces basic notations. Due to the condensed presentation of a broad spectrum of topics, it is more suitable for developing a unified viewpoint, than for an in-depth study of the subject, for which references given in the end would be useful.

We begin with a brief discussion on dynamic behavior of materials and the role of their microstructure. Both, periodic arrays of subelements and various imperfections, are found to substantially affect the material response. Then we turn to the basic concepts of the phenomenological approach, such as strain and stress, and then to constitutive laws. We deal with the linear elasticity, viscoelasticity and piezoelectricity. In fact, these physical models constitute the foundation of the linear acoustics of solids, viewed as an applied discipline.

The chapter is completed by simple considerations of basic concepts of waves, such as Debye's frequency, the signal velocity and the causality, which are of substantial interest for waves in solids. They are particularly useful for a proper understanding of the response of inhomogeneous attenuative media. The material presented raises also questions, in one form or another, concerning the dynamic response of solids and the extent to which it is affected by their properties. The chapters to follow deal with the answers to many of these questions.

1.1 Dynamic Response of Solids

There is no universal straightforward rule indicating when the dynamic nature of the phenomenon is essential, and, hence, the markedly simpler quasistatic approach has to be abandoned. The answer could be found by means of physical intuition and the analysis of a problem at hand. Consider, as an example, a steel rod of 10 cm in length subjected to a longitudinal force with frequency 10^3 Hz. It seems that the case is a typical dynamic one. However, computations reveal negligible difference in acceleration of the rod ends. It turns out, therefore, that the problem could be viewed as quasistatic. On the other hand, the same load applied to a very long rod would cause the explicit propagation of longitudinal disturbances. Due to the large dimensions of the earth even "slow" excitation is to be considered as essentially dynamic. In fact, waves due to earthquakes may have periods of 20 minutes or more.

Let L and c be a typical dimension and the wave velocity, respectively. Then, somewhat loosely speaking, one may neglect the dynamic effects if

$$L/c \ll T \quad ,$$

where T is a characteristic time interval.

The phenomenon of wave propagation in solids can be viewed as time dependent displacements, which propagate through a medium. The speed of propagation of this disturbance is finite and is referred to as the wave velocity. While an individual particle executes limited vibrations, the disturbance itself might reach any point of the medium. Consequently, the transfer of energy through space occurs. That is why the investigations of waves have practical significance.

The event of the wave propagation involves a source and a waveguide. The impacts, explosions or the evolution of material defects are well-known types of sources, while some slowly varying or moving loads may cause an explicit dynamic response as well. After emission from a source, the disturbance propagates through a waveguide, being affected by its geometrical, structural and mechanical characteristics. Due to a variety of associated phenomena, the theoretical modelling of a dynamic process may be difficult. In fact, neglect of the inertial or elastic properties of solids renders the treatment invalid. Overlooking energy dissipation, coupled electromagnetic or thermodynamic effects may also lead to incorrect conclusions. Thus, the analysis is to be based on the knowledge of the specific physical circumstances associated with the wave propagation under consideration.

1.2 Structure of Materials

Behavior of materials is governed by their atomic as well as granular structures. It was found, by means of the X-ray investigations, that in many materials, the atoms (ions) create periodic spatial arrangements, which are referred to as crystals. As Figure 1-1 displays, in this case it is always possible to determine the repeating three dimensional patterns, typical of a crystal. They all have identical environment and can be recovered from each other by integer multiples of three elementary translations defined by vectors $\boldsymbol{a}$, $\boldsymbol{b}$, and $\boldsymbol{c}$. For example, a point D is given by

$$\boldsymbol{OD} = m\boldsymbol{a} + n\boldsymbol{b} + \ell\boldsymbol{c} \quad .$$

On the microscopic scale the crystal clearly constitutes a discrete heterogeneous medium consisting of atoms and of interatomic space. However, it might be possible to ignore this discrete structure and to apply the concept of a smooth

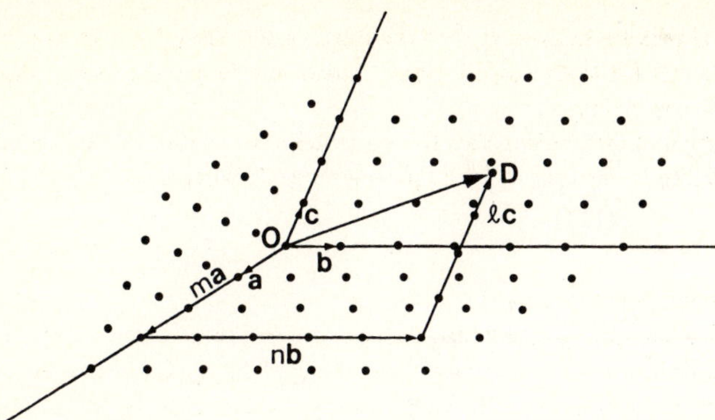

Fig. 1-1 Periodic structure of a crystal.

continuum provided the wave length is long enough. On the other hand, the periodicity of atomic arrangements gives rise to the non- equivalence of directions, that is, to anisotropy.

Due to this effect, it is necessary to differ crystal directions and planes. As Figure 1-2 shows, the direction of a ray is identified by the dimensionless coordinates of a point the ray passes through. These indices (mostly integers) are enclosed in square brackets. In order to denote the crystal plane, one exploits the points of its interception with the coordinate axes. Taking their reciprocals and enclosing them by circular brackets, one gets the so-called Miller indices of the atomic plane (see Figure 1-3).

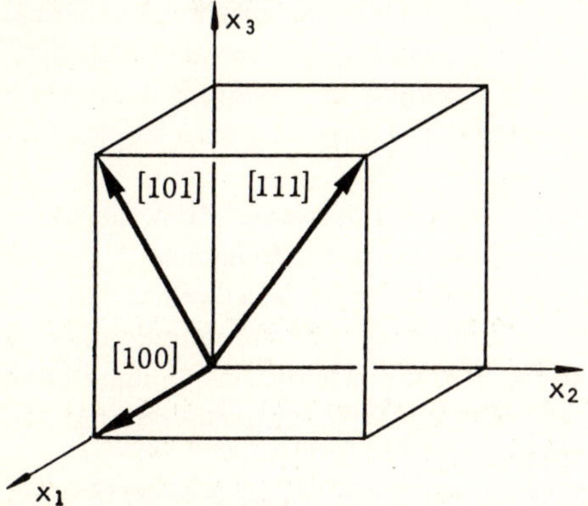

Fig. 1-2 Directions in a crystal.

The crystalline structure is mainly typical of metals. Polymers, commonly known as plastics, originate from organic raw materials and consist of giant molecules made up of repeating units (mers). The molecules are kinked and are characterized by a scatter in their size. This structure is responsible for the ability of elastomers to undergo large deformations without failure. Polymers are also particularly sensitive to temperature.

As mentioned above, monocrystals, like that of quartz shown in Figure 1-4,

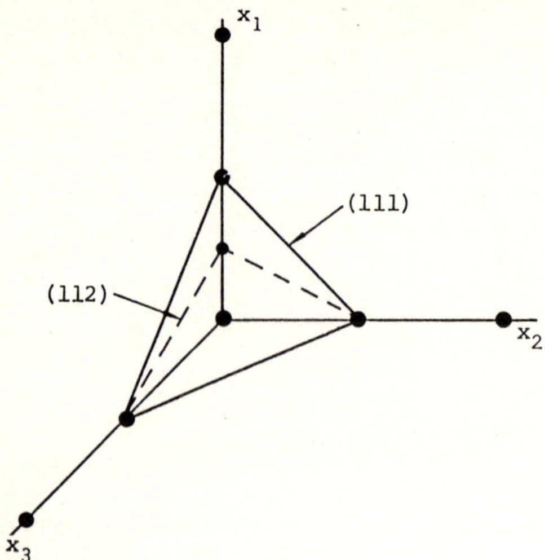

Fig. 1-3 Planes in a crystal.

exhibit an explicitly anisotropic response. Wood, laminated as well as fibrous composites, behave in a similar way. However, they are anisotropic in an overall way, unlike monocrystals, which are thought of as having a point anisotropy. On the other hand, many materials of granular structure can be treated as isotropic media. In fact, a completely chaotic orientation of grains renders equivalent all the directions. The influence of the material structure on its response is especially apparent in the case of composites. For example, chaotically dispersed spherical inclusions in a homogeneous isotropic matrix constitute the so-called "statistically isotropic" medium, since no preferable direction is present. But inserting unidirectional fibres into the same matrix produces a markedly anisotropic composite, as can be seen from Figure 1-5.

However, it would be damaging oversimplification to assume that the structure of real solids follows exactly the scheme outlined in the foregoing. A variety of deviations from a strict periodicity is typical of real crystals, as well as debonding and delamination of composites. Imperfections may have a profound influence on the dynamic behavior of materials. Moreover, the mechanisms of some basic phenomena, like plastic flow or internal friction cannot be understood unless the presence of imperfections is taken into account.

In particular, the concept of crystal dislocation has played a revolutionary role in explaining the plastic flow and strength of real solids. Figures 1-6 and 1-7 show the two basic types of dislocation, which are referred to as edge and screw, respectively. In the first case, an "extra" atomic half-plane disturbs the periodicity in the vicinity of the x_3-axis. The distortion occurs in the immediate neighbourhood of this axis, covering a space of a few lattice periods only. This region is said to be the dislocation core. In the case of a screw dislocation, shown in Figure 1-7, the crystal planes are transformed to a helicoidal surface in the vicinities of the line AB.

It is of interest that overall physical properties of the dislocations may be

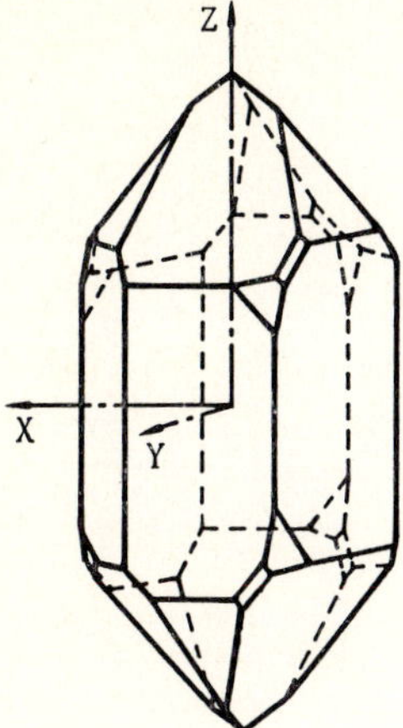

Fig. 1-4 Quartz crystal along with its crystallographic axes.

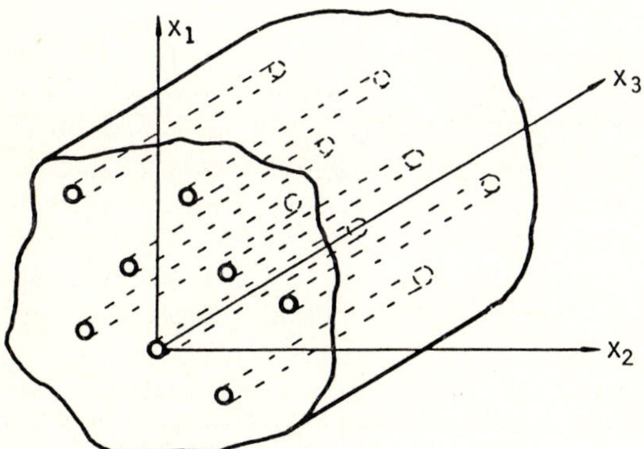

Fig. 1-5 Fiber-reinforced composite material.

described in terms of elasticity theory, which allows the displacement field, $\boldsymbol{u}$, to undergo a uniform discontinuity on some surface. A proper approach is as follows. Let us introduce a line of singularities, ℓ, of the elastic field by a unit vector $\boldsymbol{t}$ (Figure 1-8). This line is supposed to have the following property: after a passage around any closed contour, encircling the line, the displacement, $\boldsymbol{u}$, gets a finite increment $\boldsymbol{b}$. Thus

$$\int du_i = \oint \frac{du_i}{ds} ds = -b_i \neq 0 \quad ,$$

where the direction along the contour is shown in Figure 1-8. The value $\boldsymbol{b}$ is referred to as the Burgers vector. For the edge dislocation vectors $\boldsymbol{b}$ and $\boldsymbol{t}$ are mutually normal, whereas for the screw type they are parallel. In the general

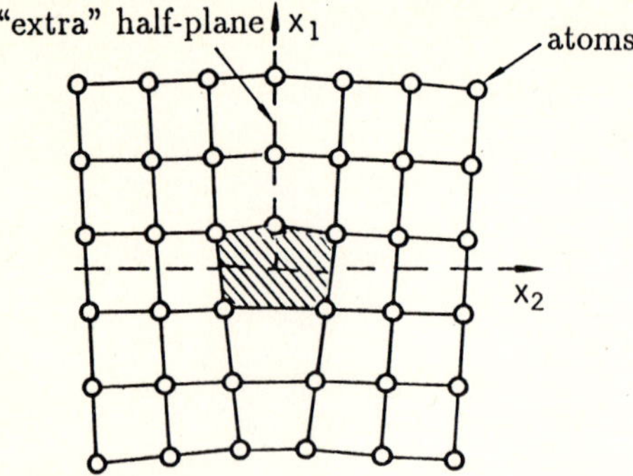

Fig. 1-6 Edge dislocation.

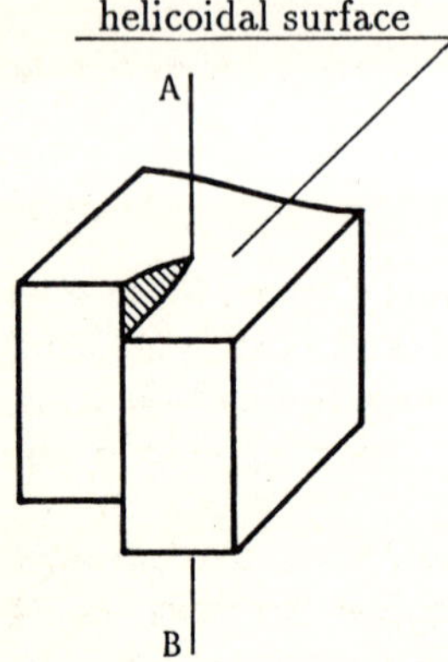

Fig. 1-7 Screw dislocation.

case b and t are neither normal nor parallel. Explicit expressions can be given for the displacement field associated with the above two types of dislocation (see Problems 1-4 and 1-5).

While the theory of elasticity was originally developed to describe purely reversible deformations, the concept of dislocation, as it has been just defined, makes it possible to explain such dissipative processes as plastic flow and internal friction.

1.3 Continuum and Microstructure

Motions of a continuum with respect to a constant coordinate reference frame are symbolically described by a function

$$X = X(x, t_1, t_2) \quad , \tag{1.1}$$

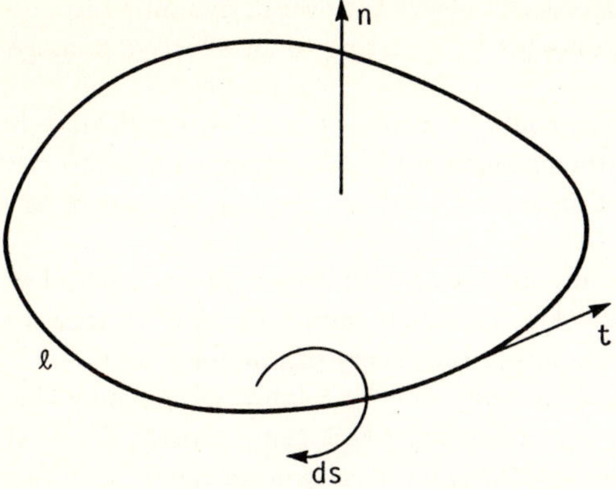

Fig. 1-8 Line of singularities describing a dislocation.

where x and X are positions of the same particle at the moments t_1 and t_2, respectively. Inversion of this equation yields

$$x = x(X, t_1, t_2) \quad . \tag{1.2}$$

The path of the particle can be followed from (1.1) provided x and t_1 are fixed while t_2 varies. This approach is referred to as Lagrangian. On the other hand, if t_1 and t_2 are fixed, (1.2) prescribes a transformation of a region of the body, $x \rightarrow X$, which occurs at the time interval $\Delta t = t_2 - t_1$. This approach to the motion of a deformable body is known as Eulerian.

To provide a clearer physical illustration, let us consider two possible ways of observing the motion of a continuum. First, one can follow the alterations of a fixed mass or a set of particles with respect to a coordinate system. Secondly, one deals with a control volume fixed relative to a reference frame and with the flux of mass, momentum and energy inside the volume as well as across its boundaries. These two examples demonstrate the Lagrangian and Eulerian approaches, respectively.

In disaccord with the concept of continuum, real materials possess explicit discrete microstructure. In general, it seems possible to deduce the mechanical response of materials by writing the equations of motion for each of their particles or subelements. This is, however, extremely cumbersome and time consuming. We revert, therefore, to the concept of continuum defined in terms of its phenomenological properties, like stresses, deformations, temperature, and construct the so-called effective homogeneous medium. It enables us to avoid the necessity of analysing all the details of particle interactions.

Consider, for example, a quartz crystal, shown in Figure 1-4 along with its crystallographic axis. The crystal consists of an explicit periodic structure with very small atoms placed at the corners of tetrahedrons. Nevertheless, when the tetrahedron dimensions are small as compared with the wavelength, the body can be fairly treated as a "standard" continuous anisotropic medium. Another example is shown in Figure 1-5 depicting a specimen of a fibrous composite material. Again, interactions among the fibers and matrix or the so-called microstructure effect must be taken into account in order to derive the overall dynamic response of this composite. This may be carried out by resorting to an effective homogeneous medium.

In general, a multiphase medium contains n phases, each of which may be treated as a continuum with its own phenomenological properties. In the case of polycrystals $n \to \infty$, since each of crystal orientations may be thought of as a different phase.

The description of an ordered composite deals with its period and with physical properties of the phases. Unlike this, a random composite identification requires a resort to quite formal techniques of the theory of random processes.

Denote a configuration of a random composite by a label, β. We may then define a probability function, $p(\beta)$, where β belongs to a sample space, U. Next, let f be a property of the composite, say, the mass density or an elastic modulus. Then its mean value is given by

$$\langle f \rangle = \int_U f(\beta) p(\beta) d\beta$$

on the understanding that in general f and p depend on a point $\boldsymbol{r}$.

Now consider the so-called indicator function, $f_q(\boldsymbol{r})$, which takes on the value 1 if $\boldsymbol{r}$ lies in phase q and zero otherwise. Then the probability of a point, $\boldsymbol{r}$, to belong to phase q is

$$P_q(\boldsymbol{r}) = \langle f_q(\boldsymbol{r}) \rangle = \int_U f_q(\boldsymbol{r}, \beta) p(\beta) d\beta \quad ,$$

while the joint probability of finding phase q at $\boldsymbol{r}$ and phase s at $\boldsymbol{r}'$ simultaneously is

$$P_{qs}(\boldsymbol{r}, \boldsymbol{r}') = \langle f_q(\boldsymbol{r}) f_s(\boldsymbol{r}') \rangle \quad .$$

Probabilities of higher order can be written in a similar way.

The above geometrical statistics enables us to obtain convenient expressions for mean values of physical parameters. Consider the case of composite with n discrete phases each having modulus μ_q. Then the variable modulus, $\mu(\boldsymbol{r})$, is

$$\mu(\boldsymbol{r}) = \sum_{q=1}^{n} \mu_q f_q(\boldsymbol{r})$$

and its two first moments are

$$\langle \mu(\boldsymbol{r}) \rangle = \sum_{q=1}^{n} \mu_q P_q(\boldsymbol{r})$$

$$\langle \mu(\boldsymbol{r}) \mu(\boldsymbol{r}') \rangle = \sum_{q=1}^{n} \sum_{s=1}^{n} \mu_q \mu_s P_{qs}(\boldsymbol{r}, \boldsymbol{r}') \quad .$$

The last expression is also called a correlation function.

A more specific description follows for a composite consisting of a matrix and dispersed identical inclusions. Then any particular configuration (specimen) can be identified by the set of $\{\boldsymbol{r}_i,\ i = 1, 2, ...\}$, where $\boldsymbol{r}_i$ is the "center" of inclusion

i. On letting P_i be a probability density for finding an inclusion centered at r_i we get

$$\int_V P_i dr_i = N \quad ,$$

where N is the number of inclusions in the volume V of the composite. Similarly, the joint probability P_{ij} for finding distinct inclusions with the centers at r_i and r_j can be defined.

Now we may introduce the notion of statistical homogeneity. This takes place when V is large compared to a microstructural spacing, like a grain size, and when the above and higher order probability densities do not depend on translations, say, P_i = const and $P_{ij} = P_{ij}(r_i - r_j)$.

The ergodic assumption applies to statistically homogeneous media. It mainly states that the average of a field variable over a large volume V and the ensemble average are the same.

1.4 Deformations

The deformation tensor describes in a quantitative manner the geometrical alterations of a continuum. It would be useful to begin with a simple one- dimensional case. Let a fiber AB (Figure 1-9) of the initial length ℓ_o undergo the deformation along the x-axis until it gets the length ℓ. The value, ε, defined as

$$\varepsilon = (\ell - \ell_o)/\ell_o \quad , \tag{1.3}$$

is an apparent quantitative characteristic of this experiment. This value is referred to as the relative elongation or the Cauchy strain measure.

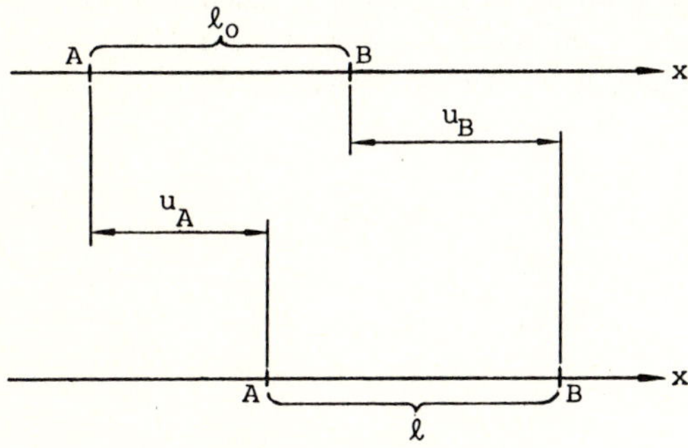

Fig. 1-9 One-dimensional displacement field.

Other measures of deformations can be constructed as well. The generalized strain measure, ε_g, is given by

$$\varepsilon_g = \int_{\ell_o}^{\ell} (\ell_o/\ell)^{n+1} d\ell/\ell_o \quad ,$$

where n is a governing parameter. For $n = -1$, the relation above yields the measure of Cauchy, given by (1.3), whereas, substituting $n = 0$ and $n = 2$, one recovers the Henky measure, ε_H, and the Almansi measure, ε_A, respectively

$$\varepsilon_H = \varepsilon_g(n = 0) = \ln(\ell/\ell_o)$$

$$\varepsilon_A = \varepsilon_g(n = 2) = \frac{1}{2}[1 - (\ell_o/\ell)^2] \quad .$$

It can be shown that under the condition of small deformations, given by

$$\Delta\ell/\ell_o = (\ell - \ell_o)/\ell_o \to o$$

the measures reduce to (1.3).

To specify the results for very small "particles" let us consider a very short fiber, setting

$$\ell_o = \Delta x \quad .$$

The displacement u_A and u_B are now related by a Taylor expansion

$$u_B = u_A + \Delta x \partial u_A/\partial x \tag{1.4}$$

and, hence,

$$\varepsilon = (u_B - u_A)/\Delta x \simeq \partial u/\partial x \quad .$$

Since A is an arbitrary point of the axis, x, and the process can be considered as time-dependent, one has $\varepsilon = \varepsilon(x, t)$.

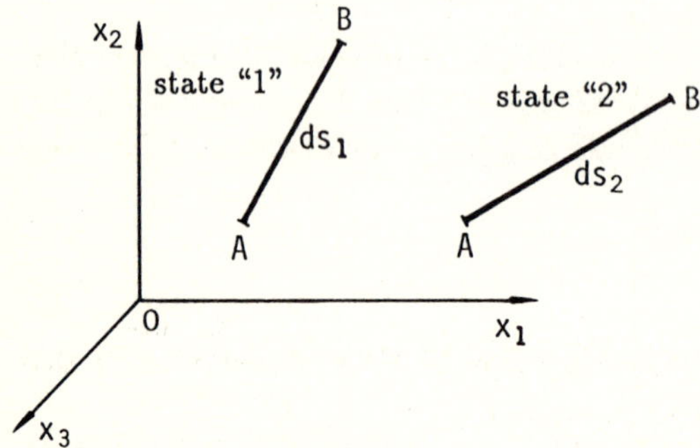

Fig. 1-10 Two states of a fiber.

Now we consider the general treatment, taking into account all three dimensions of the space. Figure 1-10 shows two different states, denoted as "1" and "2", of the same fiber. Its lengths are taken to be small and are denoted by ds_1 and ds_2, respectively. If dx_i are the projections of the fiber in the state "1" and dX_i in the state "2", then

$$(ds_1)^2 = dx_i dx_i, \qquad (ds_2)^2 = dX_i dX_i \quad .$$

Denoting the displacement along the axis x_i as u_i, one gets

$$dX_i = dx_i + du_i = dx_i + u_{i,\kappa} dx_\kappa$$

and, hence,

$$(ds_2)^2 = dx_i dx_i + 2u_{i,\kappa} dx_\kappa dx_i + u_{i,\kappa} u_{i,j} dx_\kappa dx_j \quad .$$

The increment in the square of the length is thus given by

$$(ds_2)^2 - (ds_1)^2 = 2u_{i,\kappa} dx_\kappa dx_i + u_{i,\kappa} u_{i,j} dx_\kappa dx_j \quad . \tag{1.5}$$

The symmetry of the dummy indices in the first term of the right-hand side of this equation suggests that

$$2u_{i,\kappa} dx_\kappa dx_i = u_{i,\kappa} dx_\kappa dx_i + u_{\kappa,i} dx_i dx_\kappa \quad .$$

Interchanging j and i in the last term of (1.5), one gets

$$(ds_2)^2 - (ds_1)^2 = 2e_{i\kappa} dx_i dx_\kappa \quad , \tag{1.6}$$

where the values of $e_{i\kappa}$, known as the deformation tensor, are given by

$$e_{i\kappa} = \frac{1}{2}(u_{i,\kappa} + u_{\kappa,i} + u_{j,i} u_{j,\kappa}), \quad i, \kappa, \ell = 1, 2, 3 \quad . \tag{1.7}$$

In particular, say for e_{11}, (1.7) yields

$$e_{11} = u_{1,1} + \frac{1}{2}(u_{1,1}^2 + u_{2,1}^2 + u_{3,1}^2) \quad . \tag{1.8}$$

The symmetry of the indices κ and i in (1.7) indicates that $e_{i\kappa} = e_{\kappa i}$ and a possibility of diagonalizing this tensor.

The six values, $e_{i\kappa}$, given by (1.7), enable one to determine the relative elongation along any direction as well as the change of angles during the deformation. Neglecting the square term in this equation, $|u_{j,i}| < 1$, one arrives at the linear tensor of deformation, given by

$$\varepsilon_{ij} = \frac{1}{2}(u_{i,j} + u_{j,i}) \quad . \tag{1.9}$$

Clearly, $e_{i\kappa}$ or ε_{ij} are time- and space-dependent functions, whenever a body is subjected to dynamic loading.

During deformation a particle undergoes not only strains but also rotations. The latter phenomenon is described by the antisymmetric tensor, ω_{ij}, which is, in the linear case, given by

$$\omega_{ij} = \frac{1}{2}(u_{i,j} - u_{j,i}) \quad . \tag{1.10}$$

There are only three nonvanishing components of ω_{ij}, which can be conveniently written as

$$\omega_i = -\frac{1}{2} e_{ijk} u_{k,j} \quad , \tag{1.11}$$

where e_{ijk} has the value +1, when the indices are in cyclic order, (123, 231, 312), − 1 when in acyclic order, (321, 213, 132), and zero otherwise. Thus, vector ω can be defined by

$$\omega_1 = \omega_{23}, \qquad \omega_2 = \omega_{31}, \qquad \omega_3 = \omega_{12} \quad . \tag{1.12}$$

For example, ω_1 represents the rigid-body rotation about the x_1-axis.

The six values of the deformation tensor, ε_{ij}, follow from the three functions of displacement u_i. Clearly, they cannot be completely independent. In fact, the six strains are to satisfy the compatibility conditions

$$\varepsilon_{ij,\kappa\ell} + \varepsilon_{\kappa\ell,ij} - \varepsilon_{i\kappa,j\ell} - \varepsilon_{j\ell,i\kappa} = 0$$

in order to ensure the existence of single-valued continuous functions u_i. Nevertheless, the solutions in dynamic problems are usually found directly in terms of displacements without integrating (1.9).

The components of the deformation tensor depend on an adopted coordinate system. There are, however, certain functions of them, the so-called invariants, which preserve the values in any reference frame. For example,

$$J_1 = e_{ii}, \qquad J_2 = e_{ij}e_{ji}, \qquad J_3 = e_{ij}e_{j\kappa}e_{\kappa i} \quad . \tag{1.13}$$

1.5 Stresses

Motions of a deformable body are associated with forces. The forces, applied and proportional to the volume of the body, are referred to as body forces. They are due, for example, to the presence of gravitational or electromagnetic fields and are usually long-range. On the other hand, the surface forces are exerted on a boundary via direct contact.

The concept of stresses arises from the following consideration. Let us suppose that a small surface area, ΔS, with a unit normal $\boldsymbol{n}$ is subjected to the resultant vector of surface forces $\Delta \boldsymbol{P}$. Now one defines a stress as follows:

$$\sigma_n \boldsymbol{i} = d\boldsymbol{P}/dS = \lim_{\Delta S \to 0} \Delta \boldsymbol{P}/\Delta S \quad ,$$

where $\boldsymbol{i} = \boldsymbol{P}/P$. The stress, $\sigma_n \boldsymbol{i}$, depends upon orientation, $\boldsymbol{n}$, via the relation

$$\sigma_j = \sigma_{ij} n_i \quad , \tag{1.14}$$

where σ_i, n_i are projections of $\sigma_n \boldsymbol{i}$ and $\boldsymbol{n}$, respectively, and σ_{ij} is the stress tensor. It is possible to show that $\sigma_{ij} = \sigma_{ji}$ and, thus, the tensor contains six independent components, like the tensor of deformation, ε_{ij}.

Equation (1.14) indicates that the stress, acting on any surface identified by its unit normal, $\boldsymbol{n}$, can be readily found, if the tensor σ_{ij} is known. To determine, for example, the stress acting on the plane, normal to x_1, one sets

$$n_1 = 1, \qquad n_2 = n_3 = 0$$

and making use of (1.14), gets

$$\sigma_1 = \sigma_{11}, \qquad \sigma_2 = \sigma_{12}, \qquad \sigma_3 = \sigma_{13} \quad .$$

Accordingly, the σ_{ij} can be viewed as time-dependent forces acting upon a unit cube (Figure 1-11)). This representation can be exploited to deduce the equations of motion, while the above figure is thought of as a free-body diagram. By summing up the forces acting with respect to the three coordinate directions, one gets

$$\sigma_{ij,j} + \rho f_i = \rho \ddot{u}_i \quad , \tag{1.15}$$

where ρ is the mass density and f is the resultant of body forces per unit mass.

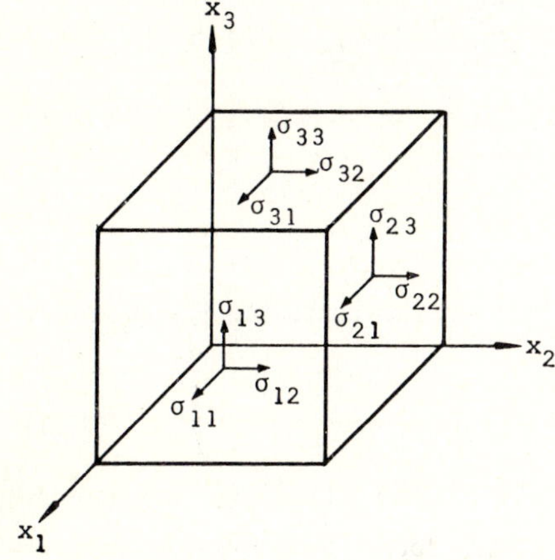

Fig. 1-11 Stresses acting on a coordinate cube.

The values of σ_{ij} depend, of course, on a reference frame. By an appropriate choice of coordinates, σ_{ij} with $i \neq j$ can be transformed to zero. This specific system of coordinates is referred to as principal. The independent invariants of the stress tensor may be written as

$$I_1 = \sigma_{ii}, \qquad I_2 = \sigma_{ij}\sigma_{ij}, \qquad I_3 = \sigma_{ij}\sigma_{j\kappa}\sigma_{\kappa i} \quad . \tag{1.16}$$

Although the tensors, σ_{ij} and e_{ij}, have many similar properties, there is the following essential difference: tensor e_{ij} is associated with the initial state of a body, whereas σ_{ij} with the final state. However, if the deformations are small, $e_{ij} \rightarrow \varepsilon_{ij}$, this difference can be neglected in order to simplify the analysis. Furthermore, the above conditions help to develop a comprehensive linear theory of waves propagation.

1.6 Conservation Laws

To derive fundamental interrelations among the basic phenomonological variables it is useful to begin with an auxiliary theorem. At time t we consider a volume

V of a moving medium bounded by a surface Σ. An external unit normal to Σ is denoted by n. During the time interval, Δt, the volume and the surface are transformed to V^* and Σ^*, respectively. As Figure 1-12

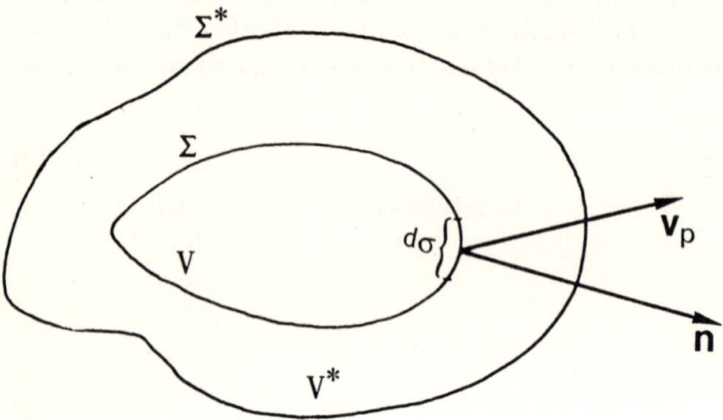

Fig. 1-12 Moving surface.

shows, the increment $\Delta V = V^* - V$ can be viewed as consisting of elementary cylinders. This gives

$$dv = v_n \Delta t d\sigma \quad , \tag{1.17}$$

where $v_n = v_i n_i$ is the normal projection of a particle velocity, v_p, and enables one to compute the value

$$\frac{d}{dt} \int_{V(t)} f(r,t) dv \quad ,$$

where $f(r,t)$ is a function of the coordinates and time, and the moveable domain, $V(t)$, depends on time.

In fact,

$$\frac{d}{dt} \int_{V(t)} f(r,t) dv = \lim_{\Delta t \to 0} \frac{\int_{V^*} f(r,t+\Delta t) dv - \int_V f(r,t) dv)}{\Delta t}$$

$$= \lim_{\Delta t \to 0} \frac{\int_V [f(r,t+\Delta t) - f(r,t)] dv + \int_{\Delta V} f(r,t+\Delta t) dv}{\Delta t} \quad .$$

The first limit in the right-hand side of this relation may be identified as

$$\frac{\partial}{\partial t} \int_V f(r,t) dv \quad ,$$

while the second one, by virtue of (1.17), as

$$\int_\Sigma f(r,t) v_n d\sigma \quad .$$

Thus,

$$\frac{d}{dt}\int_{V(t)} f(r,t)dv = \frac{\partial}{\partial t}\int_V f(r,t)dv + \int_\Sigma f(r,t)v_n d\sigma \quad . \tag{1.18}$$

In words, the total rate of change is equal to the rate of change inside the fixed volume plus the net efflux rate from the surface.

Let f be an extensive property of a set of particles and f_o the associated intensive property.

$$f = \int_m f_o dm = \int_V f_o \rho dv \quad , \tag{1.19}$$

where m and ρ are the mass and density, respectively. Then, because of (1.18), one gets

$$\frac{df}{dt} = \frac{\partial}{\partial t}\int_V f_o \rho dv + \int_\Sigma f_o \rho v_n d\sigma \quad . \tag{1.20}$$

To formulate the law of conservation of mass we define a constant M as follows:

$$M = \int_V \rho dv \quad .$$

Setting $f = M$, one gets from (1.19) that $f_o = 1$ and from (1.20)

$$\int_\Sigma \rho v_n d\sigma = \int_\Sigma \rho v_i n_i d\sigma = -\frac{\partial}{\partial t}\int_V \rho dv \quad , \tag{1.21}$$

since $dM/dt = 0$. Hence, the net efflux rate of mass from the fixed (control) surface equals the rate of change inside the volume.

Next, define the linear momentum, P_i, the resultant of body forces, F_i, and that of surface forces, Q_i, by means of the relations

$$P_i = \int_V v_i \rho dv \tag{1.22}$$

$$F_i = \int_V \rho f_i dv \tag{1.23}$$

$$Q_i = \int_\Sigma \sigma_{ij} n_j d\sigma \quad . \tag{1.24}$$

According to Newton's second law

$$dP_i/dt = F_i + Q_i \quad , \tag{1.25}$$

which is another form of (1.15). Setting $P_i = f$, we get from (1.22) and (1.19) that $v_i = f_o$ is the intensive property. Equations (1.20) and (1.25) provide then the conservation of the linear momentum

$$F_i + Q_i = \frac{\partial}{\partial t}\int_V \rho v_i dv + \int_\Sigma \rho v_i v_n d\sigma \quad . \tag{1.26}$$

The conservation theorems for the angular momentum or for the energy may be derived in a similar way.

To predict the response of a particular body under stipulated conditions, one must define, besides general expressions of the above type, the stress-strain relations $\sigma_{ij} = \sigma_{ij}(e_{ij})$, some of which are concerned with in the following sections.

1.7 Hooke's Law

It has been discovered that the behavior of many metals, polymers, and other materials under moderate loads and temperature can be fairly described by means of a linear theory. This rare combination of circumstances explains the outstanding success of the theory based on Hooke's law, which is described in the sequel.

The linear stress-strain relations are given by

$$\sigma_{ij} = c_{ij\kappa\ell}\varepsilon_{\kappa\ell} \quad , \tag{1.27}$$

where $c_{ij\kappa\ell}$ is the tensor of elastic constants. This tensor is of the 4th rank and, hence, has $3^4 = 81$ independent constants, which is, of course, disappointing. However, one can drastically reduce this number by the following analysis.

Equation (1.27) means that there is no time-effect, and the stresses are instantaneously altered along with the strains. In such cases the material is said to possess a scalar potential function, the strain energy, W, given by

$$W = \frac{1}{2}c_{ij\kappa\ell}\varepsilon_{ij}\varepsilon_{\kappa\ell} \quad . \tag{1.28}$$

Having defined this function, one obtains the stresses as

$$\sigma_{ij} = \partial W/\partial\varepsilon_{ij} \quad , \tag{1.29}$$

which is consistent with (1.27) and (1.28). Due to the symmetry of the tensor, ε_{ij}, a permutation of indices, i and j, κ and ℓ, would have no effect on W, and, thus

$$c_{ij\kappa\ell} = c_{ji\kappa\ell} = c_{ij\ell\kappa} = c_{ji\ell\kappa} \quad . \tag{1.30}$$

On the other hand,

$$\frac{\partial\sigma_{ij}}{\partial\varepsilon_{\kappa\ell}} = \frac{\partial^2 W}{\partial\varepsilon_{ij}\partial\varepsilon_{\kappa\ell}} = \frac{\partial^2 W}{\partial\varepsilon_{\kappa\ell}\partial\varepsilon_{ij}} = \frac{\partial\sigma_{\kappa\ell}}{\partial\varepsilon_{ij}}$$

and

$$c_{ij\kappa\ell} = c_{\kappa\ell ij} \quad .$$

These expressions allow the number of independent elastic constants to be reduced from 81 to 21!!

Hooke's law may now be formulated in terms of the displacements. Invoking (1.9) we get

$$\sigma_{ij} = \frac{1}{2}c_{ij\kappa\ell}u_{\kappa,\ell} + \frac{1}{2}c_{ij\kappa\ell}u_{\ell,\kappa} \quad ,$$

which in view of (1.30), is

$$\sigma_{ij} = \frac{1}{2}c_{ij\kappa\ell}u_{\kappa,\ell} + \frac{1}{2}c_{ij\ell\kappa}u_{\ell,\kappa} \quad .$$

The symmetry with respect to the dummy indices, ℓ and κ, assumes the both terms equal, which provides

$$\sigma_{ij} = c_{ij\kappa\ell}u_{\ell,\kappa} \quad . \tag{1.31}$$

Taking into account the symmetry of the tensors σ_{ij}, ε_{ij}, and $c_{ij\kappa\ell}$, one can adopt another, frequently more convenient system of labelling, which replaces two indices by one and four indices by two. The scheme of replacement is

$$\begin{array}{lll} (11) \leftrightarrow 1, & (22) \leftrightarrow 2, & (33) \leftrightarrow (3) \\ (32) = (23) \leftrightarrow 4, & (13) = (31) \leftrightarrow 5, & (21) = (12) \leftrightarrow 6 \quad . \end{array} \tag{1.32}$$

Then Hooke's law takes the matrix form with 21 elastic constants

$$\begin{Bmatrix} \sigma_1 \\ \sigma_2 \\ \sigma_3 \\ \sigma_4 \\ \sigma_5 \\ \sigma_6 \end{Bmatrix} = \begin{bmatrix} c_{11} & c_{12} & c_{13} & c_{14} & c_{15} & c_{16} \\ & c_{22} & c_{23} & c_{24} & c_{25} & c_{26} \\ & & c_{33} & c_{34} & c_{35} & c_{36} \\ & & & c_{44} & c_{45} & c_{46} \\ & & & & c_{55} & c_{56} \\ & & & & & c_{66} \end{bmatrix} \begin{Bmatrix} \varepsilon_1 \\ \varepsilon_2 \\ \varepsilon_3 \\ \varepsilon_4 \\ \varepsilon_5 \\ \varepsilon_6 \end{Bmatrix} \tag{1.33}$$

where $c_{mn} = c_{nm}$. This can be concisely written as

$$\sigma_n = c_{nm}\varepsilon_m \,, \qquad n, m = 1, 2, 3, 4, 5, 6 \quad . \tag{1.34}$$

In spite of these simplifications, the prospect of handling 21 coefficients does not seem yet attractive enough. A further progress depends on the existence of symmetry in the material structure. Consider, for example, the case when the coordinate plane x_1x_2 is a symmetry plane (mirror). It means that the transformation $x_3^1 = -x_3$, $x_2^1 = x_2$ and $x_1^1 = x_1$ causes no change in the strain energy, W, whereas the displacements are altered to

$$u_1^1 = u_1, \qquad u_2^1 = u_2, \qquad \text{and} \qquad u_3^1 = -u_3 \quad .$$

Invoking (1.9) and (1.32), one concludes that the transformations of strains are given by (in both systems of notations)

$$\begin{array}{ll} \varepsilon_{23}^1 = -\varepsilon_{23} = \varepsilon_4^1 = -\varepsilon_4 \,, & \varepsilon_{31}^1 = -\varepsilon_{31} = \varepsilon_5^1 = -\varepsilon_5 \\ \varepsilon_{11}^1 = \varepsilon_{11} = \varepsilon_1^1 - = \varepsilon_1 \,, & \varepsilon_{22}^1 = \varepsilon_{22} = \varepsilon_2^1 = \varepsilon_2 \\ \varepsilon_{33}^1 = \varepsilon_{33} = \varepsilon_3^1 = \varepsilon_3 \,, & \varepsilon_{12}^1 = \varepsilon_{12} = \varepsilon_6^1 = \varepsilon_6 \quad . \end{array} \tag{1.35}$$

The strain energy, W, is to preserve its value, which implies

$$W = \frac{1}{2}\sum_{n,m=1}^{6} c_{nm}\varepsilon_n\varepsilon_m = \frac{1}{2}\sum_{n,m=1}^{6} c_{nm}^1\varepsilon_n^1\varepsilon_m^1 \tag{1.36}$$

Substitution of (1.35) in (1.36) results in

$$c_{14} = c_{15} = c_{24} = c_{25} = c_{34} = c_{35} = c_{46} = c_{56} = 0 \quad . \tag{1.37}$$

Thus, upon existence of a plane of symmetry and the appropriate orientations of a coordinate frame, the number of independent constants reduces to 13.

A material, which has three orthogonal symmetry planes, is said to be orthotropic. The technique, similar to the foregoing, indicates that in such a case there are only 9 elastic constants. Further simplifications can be made, if the material has, along with the symmetry planes, a rotational symmetry. In particular, materials with a transverse isotropy have only 5 independent elastic constants.

In the case of cubic symmetry the elastic properties are invariant with respect to a circular permutation of the axes

$$x_1 \to x_2 \ , \qquad x_2 \to x_3 \ , \qquad x_3 \to x_1$$

and, accordingly, to a permutation of the indices

$$(123) \to (231) \to (312) \tag{1.38}$$

of the elastic constants, $c_{ij\kappa\ell}$. Hence, cubic symmetry implies that this permutation does not affect the value of the components

$$c_{1111} = c_{2222} = c_{3333} \tag{1.39}$$

$$c_{1122} = c_{2233} = c_{3311} \ , \qquad c_{1212} = c_{2323} = c_{3131} \quad .$$

Invoking (1.32), (1.33) and (1.37), one finds that only three (!!) independent elastic constants are needed in this case, c_{11}, c_{12} and c_{44}:

$$c_{nm} = \begin{bmatrix} c_{11} & c_{12} & c_{12} & 0 & 0 & 0 \\ c_{12} & c_{11} & c_{12} & 0 & 0 & 0 \\ c_{12} & c_{12} & c_{11} & 0 & 0 & 0 \\ 0 & 0 & 0 & c_{44} & 0 & 0 \\ 0 & 0 & 0 & 0 & c_{44} & 0 \\ 0 & 0 & 0 & 0 & 0 & c_{44} \end{bmatrix} \quad . \tag{1.40}$$

It should be stressed, that this effect takes place only when the reference frame coincides with the symmetry axes of the material.

For the case of isotropic media, the elastic properties are invariant with respect to any rotation of a reference frame, which imposes an additional constraint,

$$c_{44} = (c_{11} - c_{12})/2 \quad . \tag{1.41}$$

Denoting

$$c_{12} = \lambda \qquad \text{and} \qquad c_{44} = \mu \quad ,$$

one gets

$$c_{11} = \lambda + 2\mu \quad ,$$

and Hooke's law can be written as

$$\sigma_{ij} = \lambda \tilde{\theta} \delta_{ij} + 2\mu \varepsilon_{ij} \quad , \tag{1.42}$$

where $\tilde{\theta} = \varepsilon_{ii}$. The constants, λ and μ, which are referred to as the Lame moduli, closely relate to the other elastic parameters, K, G, E and ν, known as the bulk modulus, shear modulus, Young's modulus, and Poisson's ratio, respectively. Elastic deformations are accompanied by temperature changes. However, the associated heat transfer is quite slight and the process may be considered as adiabatic. Accordingly, the elastic constants appearing in the linear wave analysis are mostly the adiabatic values. Thermodynamic considerations of constitutive laws will be given in Section 1.9.

1.8 Absorption and Viscoelasticity

In the theory of elasticity the stresses and strains are related by constants. Accordingly, the time-histories of these values are similar and the deformation process is completely reversible. It was found out, however, that most polymers as well as metals, particularly, under higher temperature, exhibit an explicit departure from this type of behavior. For example, a strain (stress) changes in time, while the counterpart remains constant. Many porous and fibrous materials partially absorb the energy. These deviations from pure elasticity may be taken into account by replacing the elastic constants, which appear in Hooke's law, by integral or differential time-operators.

To illustrate this approach consider a one-dimensional case described by the stress σ and strain ε. While the purely elastic component of ε, denoted as ε_e, and the stress are again related by a constant, E,

$$\varepsilon_e = \sigma / E \quad ,$$

assumptions should be made concerning the accumulated portion of ε, ε_v, which reflects the previous history of the specimen up to the time, t. It is assumed that an increment, $d\varepsilon_v$, is proportional to the stress, $\sigma(\tau)$, the duration of its action, $d\tau$, and depends on the time interval, $(t-\tau)$, via a function $K(t-\tau)$

$$d\varepsilon_v = \frac{\sigma(\tau)}{E} d\tau K(t-\tau) \quad .$$

Integrating this relation from $-\infty$ to t and taking into account the purely elastic term one gets

$$\varepsilon = \varepsilon_e + e_v = \frac{\sigma}{E} + \frac{1}{E} \int_{-\infty}^{t} K(t-\tau)\sigma(\tau) d\tau \quad .$$

Similar considerations provide the reciprocal relation

$$\sigma = E\varepsilon + E \int_{-\infty}^{t} \Gamma(t-\tau)\varepsilon(\tau) d\tau \quad .$$

Since the kernels of the above equations, $K(t-\tau)$ and $\Gamma(t-\tau)$, depend solely on the time difference, the material at hand has no sensitivity to the translation of the time scale.

An extension of this approach to the three-dimensional case for isotropic viscoelastic media may be cast as follows:

$$\begin{aligned} s_{ij} &= \int_{-\infty}^{t} G_1(t-\tau)(d\varepsilon_{ij}^{d}(\tau)/d\tau)d\tau \\ \sigma_{kk} &= \int_{-\infty}^{t} G_2(t-\tau)(d\varepsilon_{kk}(\tau)/d\tau)d\tau \quad , \end{aligned} \tag{1.43}$$

where s_{ij} and ε_{ij}^{d}, known as deviatoric tensors, are

$$\begin{aligned} s_{ij} &= \sigma_{ij} - 1/3\delta_{ij}\sigma_{kk} \, , \qquad & s_{ii} &= 0 \\ \varepsilon_{ij}^{d} &= \varepsilon_{ij} - 1/3\delta_{ij}\varepsilon_{kk} \, , \qquad & \varepsilon_{ii}^{d} &= 0 \quad , \end{aligned} \tag{1.44}$$

and $G_1(\alpha)$ and $G_2(\alpha)$ are referred to as the relaxation functions of the material.

An alternative formulation makes use of differential operators

$$\begin{aligned} P(D)s_{ij}(t) &= Q(D)\varepsilon_{ij}^{d}(t) \\ L(D)\sigma_{kk}(t) &= M(D)\varepsilon_{kk}(t) \quad , \end{aligned} \tag{1.45}$$

where each of the operators has the form

$$F(D) = \sum_{n} C_n d^n/dt^n \quad ,$$

with C_n denoting material constants.

As in any linear theory, the harmonic case, when $\sigma_{ij} = \sigma_{ij}(\omega)$, $\varepsilon_{ij} = \varepsilon_{ij}(\omega)$, is of particular interest. By application of the Fourier transform, the above equations can be shown to provide

$$\begin{aligned} s_{ij}(\omega) &= G_1(i\omega)\varepsilon_{ij}^{d}(\omega) \\ \sigma_{kk}(\omega) &= G_2(i\omega)\varepsilon_{kk}(\omega) \quad , \end{aligned} \tag{1.46}$$

where the complex dynamic moduli are given by

$$G_\alpha(i\omega) = G_\alpha^r(\omega) + iG_\alpha^i(\omega), \qquad \alpha = 1,2 \quad .$$

Mechanical models consisting of springs and dashpots have been invoked to arrive at particular forms of these functions. These models have rather qualitative than quantitative capability.

The real and imaginary parts of the moduli are related by the so-called Kramers-Kroniq equations,

$$\begin{aligned} G_\alpha^r &= \frac{2}{\pi} P \int_0^\infty \frac{\varepsilon G_\alpha^i(\varepsilon)d\varepsilon}{\varepsilon^2 - \omega^2} \\ G_\alpha^i &= -\frac{2}{\pi} P \int_0^\infty \frac{\omega G_\alpha^r(\varepsilon)d\varepsilon}{\varepsilon^2 - \omega^2} \quad , \end{aligned} \tag{1.47}$$

where P denotes Cauchy principal value. It will be shown in Section 1.17 that the above relations follow from the causality and linearity of the model. One observes from (1.43), as well as from (1.46) and (1.47), that the viscoelastic response of an isotropic medium is defined by two real functions depending either on time or on frequency.

The mechanism responsible for the internal energy-loss is not yet clearly understood. Besides viscoelasticity, the attenuation of waves in solids is affected, among others, by thermal effects and scattering by inclusions, full account of which is difficult. It seems that relative relevance of these factors depends on the range of frequencies and the material structure. Several characteristics are used to describe the overall ability of a material to absorb mechanical energy. One of them, the logarithmic decrement, δ, is defined for the harmonic vibrations of a sample and is given by

$$\delta \simeq \frac{\Delta W}{2W_s} \quad , \tag{1.48}$$

where ΔW is the energy-loss per cycle and W_s is the total energy of the specimen. By analogy with the theory of electric circuits, the quality factor, Q, is

$$Q = \pi/\delta \quad . \tag{1.49}$$

The correspondence principle enables one to obtain the solution to a viscoelastic problem from the associated purely elastic one. In the case of harmonic waves its application is especially simple: replace the elastic constants by the appropriate viscoelastic moduli. This operation is valid under the following restrictions: i) no boundary conditions depend explicitly on time and ii) the elastic solution has been obtained without separating imaginary and real parts and without determining the modul of a complex value.

The procedure is much more complicated in the case of transient phenomena. Prior to the replacement it is necessary to apply the Fourier or Laplace transform to the basic relations of the associated elastic problem. Then, the inverse transform (which, however, may turn out difficult to carry out) would provide the solution.

It should be pointed out that the complete treatment of viscoelastic effects necessitates, strictly speaking, a thermodynamic analysis, since the energy-loss may lead to a significant heat production within the material. The approach, outlined in the foregoing, applies only to low-loss and structurally stable materials.

1.9 Piezoelectricity and Other Coupled Phenomena

Some crystalline materials as well as ceramics and polymers may exhibit the coupling between electrical and mechanical responses, that is referred to as piezoelectricity. The lack of symmetry in location of electrical charges, associated with the ions, is thought of as responsible for this phenomenon.

The basic mechanism of piezoelectricity can be demonstrated in a simple way by means of Figure 1-13 depicting, say, the cadmium and sulphur ions with the electrical charges $(-q)$ and $(+q)$, respectively. The electrical forces due to these charges are represented by the two springs. To reflect the lack of symmetry the

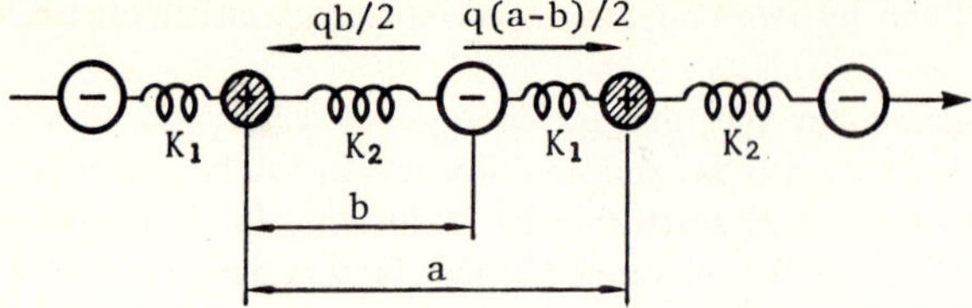

Fig. 1-13 One-dimensional simplified model of piezoelectricity.

springs are supposed to have different stiffnesses, say, K_1 and K_2 ($K_1 \neq K_2$). Accordingly, the displacements of the ions are given by a and b with $a \neq 2b$. The dipole moments are of the opposite directions and have the magnitudes $q(a-b)/2$ and $qb/2$. The resultant (electrical polarisation) is

$$P = q(a-2b)/2 \quad .$$

Superimposing an external axial stress on the chain produces the increments, Δa and Δb, and the increment in polarisation of a molecule given by

$$\Delta P = q(\Delta a - 2\Delta b)/2 \quad .$$

This is a simplified illustration of conversion of the mechanical energy to electrical, known as the direct piezoelectric effect. On the other hand, one can easily imagine that subjecting the chain to an external electric field would give rise to the ions displacements, resulting in deformations of the springs and elastic forces. This is the inverse piezoelectric effect.

We turn to a general case, involving the electric, magnetic, and temperature effects. Thermodynamic analysis provides a convenient means for derivation of the constitutive equations for this case. The system is described by the following variables: i) tensors of strains and stresses, ε_{ij} and σ_{ij}; ii) electric field, $\boldsymbol{E}$, and electric displacement, $\boldsymbol{D}$; iii) magnetic field, $\boldsymbol{H}$, and magnetic induction, $\boldsymbol{B}$; and iv) temperature, T, and entropy, S.

Selecting, for example, σ_{ij}, E_i, H_i, and T as independent, and ε_{ij}, D_i, B_i, and S as dependent variables, one may write

$$d\varepsilon_{ij} = \left(\frac{\partial \varepsilon_{ij}}{\partial \sigma_{k\ell}}\right)_{EHT} d\sigma_{k\ell} + \left(\frac{\partial \varepsilon_{ij}}{\partial E_k}\right)_{\sigma HT} dE_k + \left(\frac{\partial \varepsilon_{ij}}{\partial H_k}\right)_{\sigma ET} dH_k + \left(\frac{\partial \varepsilon_{ij}}{\partial T}\right)_{\sigma EH} dT,$$

$$dD_i = \left(\frac{\partial D_i}{\partial \sigma_{jk}}\right)_{EHT} d\sigma_{jk} + \left(\frac{\partial D_i}{\partial E_j}\right)_{\sigma HT} dE_j + \left(\frac{\partial D_i}{\partial H_j}\right)_{\sigma ET} dH_j + \left(\frac{\partial D_i}{\partial T}\right)_{\sigma EH} dT,$$

$$dB_i = \left(\frac{\partial B_i}{\partial \sigma_{jk}}\right)_{EHT} d\sigma_{jk} + \left(\frac{\partial B_i}{\partial E_j}\right)_{\sigma HT} dE_j + \left(\frac{\partial B_i}{\partial H_j}\right)_{\sigma ET} dH_j + \left(\frac{\partial B_i}{\partial T}\right)_{\sigma EH} dT,$$

$$dS = \left(\frac{\partial S}{\partial \sigma_{ij}}\right)_{EHT} d\sigma_{ij} + \left(\frac{\partial S}{\partial E_i}\right)_{\sigma HT} dE_i + \left(\frac{\partial S}{\partial H_i}\right)_{\sigma ET} dH_i + \left(\frac{\partial S}{\partial T}\right)_{\sigma EH} dT. \tag{1.50}$$

In the above, $(\partial f/\partial x)_y$ means that the derivative is taken while y is a constant. According to the first law of thermodynamics the internal energy of the system, U is governed by

$$dU = \sigma_{ij} d\varepsilon_{ij} + E_i dD_i + H_i dB_i + T dS \quad , \tag{1.51}$$

where the first three terms in the right-hand side are, respectively, the elastic, electric and magnetic energies, while the last term is the heat. It is convenient to introduce Gibbs potential, G, by means of the relation

$$G = U - \sigma_{ij}\varepsilon_{ij} - E_i D_i - H_i B_i - TS \quad .$$

The differential, dG, is then given by

$$dG = -\varepsilon_{ij} d\sigma_{ij} - D_i dE_i - B_i dH_i - S dT \quad , \tag{1.52}$$

where (1.51) has been employed. Thus, G is a function of σ_{ij}, E_i, H_i, and T, and one may write

$$dG = \left(\frac{\partial G}{\partial \sigma_{ij}}\right)_{EHT} d\sigma_{ij} + \left(\frac{\partial G}{\partial E_i}\right)_{\sigma HT} dE + \left(\frac{\partial G}{\partial H_i}\right)_{\sigma ET} dH + \left(\frac{\partial G}{\partial T}\right)_{\sigma EH} dT \quad . \tag{1.53}$$

Comparing (1.52) and (1.53) one gets

$$\varepsilon_{ij} = -\left(\frac{\partial G}{\partial \sigma_{ij}}\right)_{EHT}, \quad D_i = -\left(\frac{\partial G}{\partial E_i}\right)_{\sigma HT}$$
$$B_i = -\left(\frac{\partial G}{\partial H}\right)_{\sigma HT}, \quad S = -\left(\frac{\partial G}{\partial T}\right)_{\sigma EH} \quad . \tag{1.54}$$

These equations are extremely useful. In fact, by the straightforward differentiation it is now possible to relate the derivatives appearing in (1.50). For example,

$$\left(\frac{\partial \varepsilon_{ij}}{\partial E_k}\right)_{\sigma HT} = \left(\frac{\partial D_k}{\partial \sigma_{ij}}\right)_{EHT} = d^{TH}_{kij} \quad .$$

This and other derivatives of interest are given in Table 1.1 where α denotes an independent variable, while β its function, $\frac{\partial\beta}{\partial\alpha}$. For example,

$$n_{ki}^{\sigma T} = \left(\frac{\partial B_i}{\partial E_k}\right)_{\sigma T} \quad .$$

Table 1.1 Notation of derivatives in (1.55)

α/β	ε_{ij}	D_i	B_i	S
$\sigma_{k\ell}$	$s_{ijk\ell}^{EHT}$	$d_{ik\ell}^{TH}$	$d_{ik\ell}^{TE}$	$\alpha_{k\ell}^{EH}$
E_k	d_{kij}^{TH}	$\kappa_{ik}^{\sigma HT}$	$n_{ki}^{\sigma T}$	$p_k^{\sigma H}$
H_k	d_{kij}^{TE}	$n_{ik}^{\sigma T}$	$\mu_{ik}^{\sigma EH}$	$i_k^{\sigma E}$
T	α_{ij}^{EH}	$p_i^{\sigma H}$	$i_i^{\sigma E}$	$C^{\sigma EH}/T$

One observes that the matrix of the coefficients is symmetric. Adopting these notations and making use of (1.50), one arrives at the following constitutive relations

$$\begin{aligned} \varepsilon_{ij} &= s_{ijk\ell}^{EHT}\sigma_{k\ell} + d_{kji}^{HT}E_k + d_{kij}^{ET}H_k + \alpha_{ij}^{EH}\Delta T \\ D_i &= d_{ijk}^{HT}\sigma_{jk} + \kappa_{ij}^{\sigma HT}E_j + n_{ij}^{\sigma T}H_j + p_i^{\sigma H}\Delta T \\ B_i &= d_{ijk}^{ET}\sigma_{jk} + n_{ji}^{\sigma HT}E_j + \mu_{ij}^{\sigma ET}H_j + i_i^{\sigma E}\Delta T \\ \Delta S &= \alpha_{ij}^{EH}\sigma_{ij} + P_i^{\sigma H}E_i + i_i^{\sigma E}H_i + C^{\sigma EH}\Delta T/T \quad . \end{aligned} \tag{1.55}$$

From a different choice of independent and dependent variables, alternative forms of equations can be deduced. It should be pointed out that either the electric or the magnetic effect is usually predominant in its influence on the elastic field. Equations (1.55) admit then appropriate simplifications.

One observes from these relations that the elastic, piezoelectric, and dielectric coefficients are defined for a constant temperature. Nevertheless, the adiabatic (S = const) and the isothermal (T = const) coefficients can be interrelated by a similar approach.

We shall be concerned with piezoelectric materials under constant temperature, neglecting the magnetic effects. The equations for this particular case are given by

$$D_i = \zeta_{ij}^{\varepsilon}E_j + f_{ijk}\varepsilon_{jk} \tag{1.56}$$

$$\sigma_{ik} = c_{jk\ell m}^{E}\varepsilon_{\ell m} - f_{ijk}E_i \quad , \tag{1.57}$$

where new notations have been used*, namely, f_{ijk} is the tensor of piezoelectric constants and ζ_{ij} is that of dielectric constants. In the convenient matrix notation

* For standard notations see IEEE Standard on Piezoelectricity (IEET Trans., Sonics and Ultrasonics, Vol. SU-31, No.2, 1984).

introduced in Section 1.7, these relations take the form

$$D_i = \zeta^{\varepsilon}_{ij} E_j + f_{i\alpha}\varepsilon_\alpha \tag{1.58}$$

$$\sigma_\alpha = c^{E}_{\alpha\beta}\varepsilon_\beta - f_{i\alpha}E_i \tag{1.59}$$

$$(i,j = 1,2,3, \quad \alpha,\beta = 1,2....6) \quad .$$

Here the superscripts E and ε indicate that the value is defined for E = const or ε_{ij} = const, respectively. In fact, due to electromechanical coupling, one has to specify exactly the conditions under which the law is valid.

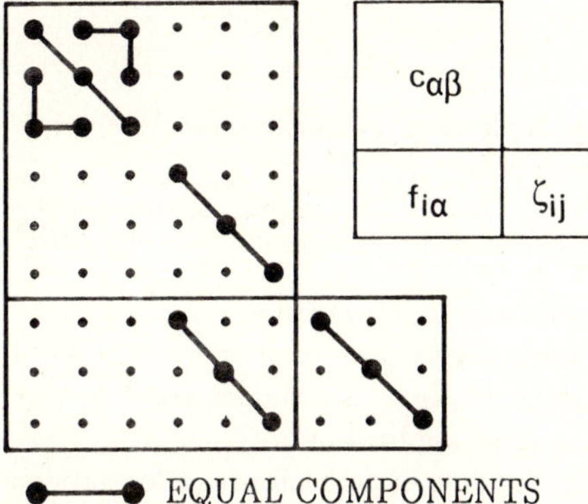

Fig. 1-14 Matrices for elastic, piezoelectric, and dielectric constants in the crystallographic axes.

Reduction in the number of independent coefficients can be carried out by crystal symmetry, similarly to the procedure of Section 1.7. Figure 1-14 shows the matrices for elastic, piezoelectric, and dielectric constants for a cubic system in the crystallographic axes reference frame. The layout is explained in the upper corner of the figure (see also (1.40)).

1.10 Variational Theorems

In the statics of elastic solids two fundamental energy principles are of extreme usefulness. One of them, known as the theorem of minimum potential energy, states that of all displacements, u_i, satisfying the given boundary conditions, those, which satisfy the equilibrium equations, are distinguished by an absolute minimum of the potential energy functional defined by

$$U_\varepsilon = \int_V [W(\varepsilon_{ij}) - f_i u_i] dv - \int_{S_\sigma} q_i u_i ds \quad .$$

Here S_σ is a portion of the surface of the body of volume V with prescribed tractions q_i, and W is given by (1.28). Denoting this minimum as U_ε^{ex} we may write

$$U_\varepsilon > U_\varepsilon^{ex} \quad , \tag{1.60}$$

where U_ε is the functional evaluated for any admissible displacements.

The other relevant result, referred to as the theorem of minimum of complementary energy, states that of all stresses, σ_{ij}, satisfying the equilibrium equations and the stress boundary conditions, those, which satisfy the compatibility equations, are distinguished by an absolute minimum of the complementary energy functional, defined by

$$U_\sigma = \int_V W(\sigma_{ij})dv - \int_{S_u} \sigma_i u_i ds \quad .$$

Here S_u is a portion of the surface of the body with prescribed displacements, u_i. Denoting this minimum as U_σ^{ex}, we may write

$$U_\sigma > U_\sigma^{ex} \quad , \tag{1.61}$$

where U_σ is the functional evaluated for any admissible stresses.

Mixed functionals, in which both, stresses and strains, may be subjected to variations, have also been derived. However, they are characterized by the condition of stationarity only, which limits their application.

The basic variational formulation for time-dependent elastic deformations is known as Hamilton's principle. To this end, consider virtual displacements, $\delta u_i(\boldsymbol{r}, t)$, which are consistent with the boundary conditions prescribed on S_u and are "smooth" but otherwise arbitrary. Then the work done by the body forces, f_i, and surface tractions, q_i, is

$$\delta A = \int_V f_i \delta u_i dv + \int_{S_\sigma} q_i \delta u_i ds \quad . \tag{1.62}$$

The surface integral may be transformed as follows:

$$\int_{S_\sigma} q_i \delta u_i ds = \int_V (\sigma_{ij}\delta u_i)_{,j}\, dv = \int_V (\sigma_{ij,j}\delta u_i + \sigma_{ij}\delta u_{i,j})dv \quad ,$$

where use has been made of the relation $q_i = \sigma_{ij} n_j$ and Gauss' theorem. It can be shown that $u_{i,j}$ splits up into symmetric and antisymmetric parts

$$\delta u_{i,j} = \delta\varepsilon_{ij} + \delta\omega_{ij} \quad ,$$

which, in view of symmetry of σ_{ij} and (1.10), implies

$$\sigma_{ij}\delta u_{i,j} = \sigma_{ij}\delta\varepsilon_{ij} \quad .$$

Hence,

$$\delta A = \int_V (\rho \ddot{u}_i \delta u_i + \sigma_{ij}\delta\varepsilon_{ij})dv \quad , \tag{1.63}$$

where $f_i = \rho\ddot{u}_i$ has been employed. We may recognize that the first term should be associated with the total kinetic energy, K_{tot}, given by

$$K_{tot} = \frac{1}{2}\int_V \rho\dot{u}_i\dot{u}_i dv \quad ,$$

while the second with the total strain energy, W_{tot},

$$W_{tot} = \frac{1}{2}\int_V \sigma_{ij}\varepsilon_{ij} dv = \int_V W dv \quad .$$

In fact, on integrating with respect to time and introducing abbreviation

$$I = \int_{t_0}^{t_1}\int_V \rho\ddot{u}_i\delta u_i dv dt$$

with t_0 and t_1 denoting two arbitrary instants, we get

$$I = \int_V \rho\dot{u}_i\delta u_i dv \Big|_{t_0}^{t_1} - \int_V dv \int_{t_0}^{t_1} \frac{\partial}{\partial t}(\rho\delta u_i)\dot{u}_i dt \quad ,$$

where integration by parts over time has been used. By imposing $\delta u_i\ (t = t_o) = \delta u_i\ (t = t_1)\ = 0$ this expression may be rewritten as

$$I = -\int_V \rho dv \int_{t_0}^{t_1} \dot{u}_i\delta\dot{u}_i dt = \int_{t_0}^{t_1} \delta K_{tot} dt \quad .$$

Similarly, for the remaining part of (1.63) we obtain

$$\int_V \sigma_{ij}\delta\varepsilon_{ij} dv = \delta W_{tot},$$

and, consequently, with the help of (1.63), we get

$$\delta\int_{t_0}^{t_1}(W_{tot}-K_{tot}-A)dt = 0 \quad , \tag{1.64}$$

which provides Hamilton's principle for a perfectly elastic linear solid. A relevant point to note is that a modification of this principle applies to dissipative systems too.

1.11 Wave Front and Wave Classification

When a wave propagates through a solid, the latter is divided into two subregions: disturbed, V_1, and undisturbed, V_2 (Figure 1-15). Denote the separating surface as $S(t)$, where the argument, t, indicates that the surface is supposed to move through the medium. This motion is called an elastic wave, if the material in each subregion is elastic. The plastic wave takes place provided the material is in a plastic state. A wave of plastic loading occurs upon the condition that a sub-volume V_2 is elastic, while V_1 plastic. These examples show that the medium

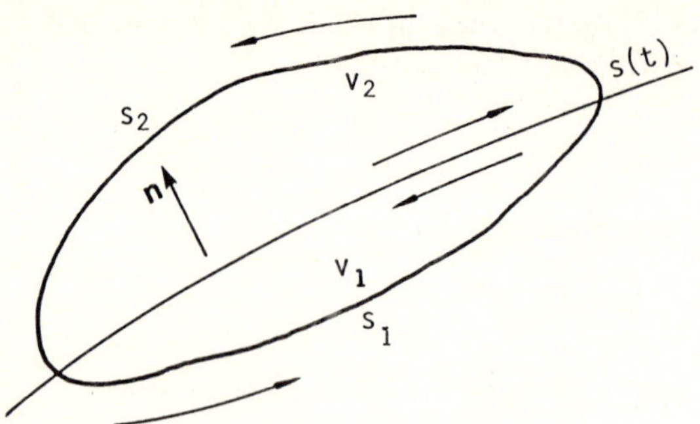

Fig. 1-15 Moving surface $S(t)$ and subregions V_1 and V_2.

may behave in different ways, depending on the situation in the sub-volumes divided by an extremely "thin" region.

Let us introduce the displacement, $\boldsymbol{u}(\boldsymbol{r},t)$. Derivatives of $\boldsymbol{u}$ with respect to x_i or t may experience jumps across the surface $S(t)$. This surface is said to be a discontinuity surface of n-th order, if the minimal order of such a derivative is n. If $n \geq 2$, the event is called a weak discontinuity wave, whereas for $n = 1$ a wave of strong discontinuity. For example, a jump in the strain rate, $\dot{\varepsilon}$, belongs to the first type, while a jump in σ_{ij} or ε_{ij} to the second. The case $n = 0$ corresponds to a discontinuity in the displacement and may constitute a dislocation (see Section 1.2).

Intending to define the velocity of wave propagation, we consider two states of the moving surface, $S(t)$, corresponding to t and $t + \Delta t$ (Figure 1-16). Assume that the equation of the surface is given by

$$F(\boldsymbol{r},t) = 0 \quad . \tag{1.65}$$

Then its unit normal-vector, $\boldsymbol{n}$, acquires the form

$$\boldsymbol{n} = \textbf{grad}\, F / |\textbf{grad}\, F| \quad . \tag{1.66}$$

Denoting the length of the normal between the surfaces as ΔL, we define the velocity of propagation, $\boldsymbol{c}$, as

$$\boldsymbol{c} = \boldsymbol{n} \lim_{\Delta t \to 0} \frac{\Delta L}{\Delta t} = \boldsymbol{n}\frac{dL}{dt} \quad . \tag{1.67}$$

To determine an explicit expression for $\boldsymbol{c}$, one recalls that the equation of the surface for $t + \Delta t$ can be written down, according to (1.65), as

$$F(x_i + \Delta L n_i,\ t + \Delta t) = 0 \quad .$$

Expanding this relation for small increments one gets

$$\Delta L \frac{\partial F}{\partial x_i} n_i + \frac{\partial F}{\partial t}\Delta t = \Delta L |\textbf{grad}\, F| + \frac{\partial F}{\partial t}\Delta t = 0 \quad ,$$

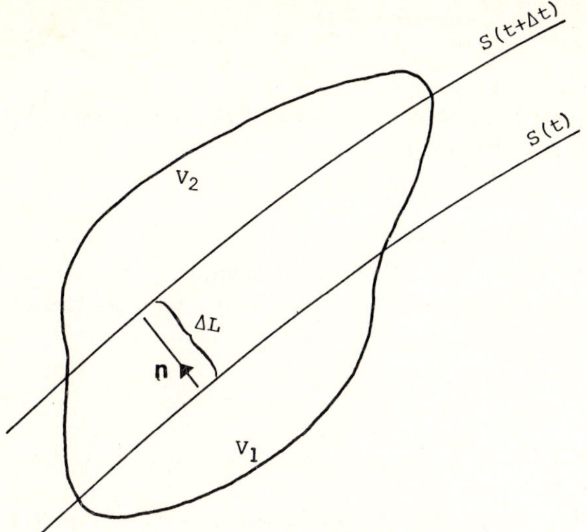

Fig. 1-16 States of a moving surface at t and $t + \Delta t$.

which results finally in

$$c = -n\frac{\partial F}{\partial t}\frac{1}{|\mathbf{grad}\, F|} \quad . \tag{1.68}$$

Thus, the velocity of propagation can be found directly from the equation of the moving surface.

1.12 Conditions Across Discontinuity

Besides the above propagation velocity, c, there is a particle velocity, $\boldsymbol{v}(\boldsymbol{r}, t)$, associated with the displacement function, $\boldsymbol{u}(\boldsymbol{r}, t)$. Is it possible to relate between them in terms of the medium parameters? What is the equation governing the jump evolution during its motion through a medium?

The first question can be answered by employing (1.18), which yields the derivative of an integral taken over a volume with moveable boundaries. Applying this result to both, V_1 and V_2, and adjusting the notations, one gets for an arbitrary "regular" function, $f(\boldsymbol{r}, t)$

$$\begin{aligned} \frac{d}{dt}\int_{V_1} f(\boldsymbol{r}, t)dv &= \int_{V_1} \frac{\partial f}{\partial t}dv + \int_{S_1} f v_n ds + \int_S f c ds \\ \frac{d}{dt}\int_{V_2} f(\boldsymbol{r}, t)dv &= \int_{V_2} \frac{\partial f}{\partial t}dv + \int_{S_2} f v_n ds - \int_S f c ds \quad , \end{aligned} \tag{1.69}$$

where the meaning of the surfaces S, S_1, and S_2 should be clear from Figure 1-15 (arrows indicate the direction employed in integration).

Assume that $f(r,t)$ is discontinuous across $S(t)$:

$$[f] = f_2 - f_1 \neq 0 \quad ,$$

where subscripts 1 and 2 denote the values on the sides of $S(t)$ belonging to V_1 and V_2, respectively. Summing (1.69), we get

$$\frac{d}{dt}\int_V f dv = \int_V \frac{\partial f}{\partial t} dv + \int_{S_1+S_2} f v_n ds - \int_S [f] c ds. \tag{1.70}$$

This relation has various applications. Define, for example, $f(r,t)$ as the medium density, $f = \rho$. Taking into account that there is no total mass change during the deformation,

$$\frac{d}{dt}\int_V \rho dv = 0 \quad ,$$

one finds from (1.70) that $\mathbf{c}$ and $\boldsymbol{v}$ are related by

$$\int_V \frac{\partial \rho}{\partial t} dv + \int_{S_1} \rho v_n ds + \int_{S_2} \rho v_n ds - \int_S [\rho] c ds = 0, \tag{1.71}$$

with

$$[\rho] = \rho_2 - \rho_1$$

being the density jump across the surface $S(t)$.

Next, one imagines that $V \to 0$ so that

$$S_1 \to S_o , \qquad S_2 \to (-S_o) \quad ,$$

where S_o is a small part of the surface, $S(t)$. Then (1.71), with the help of Figure 1-15, yields

$$\int_{S_o} \rho_1 v_{1n} ds - \int_{S_o} \rho_2 v_{2n} ds + \int_{S_o} [\rho] c ds = \int_{S_o} (\rho_1 v_{1n} - \rho_2 v_{2n} + [\rho] c) ds = 0 \quad ,$$

where v_{1n} and v_{2n} denote the particle velocities normal to the separating surface, $S(t)$, from the sides in V_1 and V_2, respectively. Due to the arbitrariness of S_o, the above relation provides a condition across the discontinuity

$$\rho_1(v_{1n} - c) = \rho_2(v_{2n} - c) \quad . \tag{1.72}$$

Other relations can be deduced by the same technique. In particular, upon substitution

$$f_1 = \rho_1 v_{1i}, \qquad f_2 = \rho_2 v_{2i}$$

and with the help of the principle of momentum conservation, given by (1.26), one may get

$$[\sigma_{ij}] n_j = \rho_1 (v_{1n} - c)[v_i] \quad , \tag{1.73}$$

where $[\sigma_{ij}] = \sigma_{2ij} - \sigma_{1ij}$, $[v_i] = v_{2i} - v_{1i}$. The above relation is referred to as the dynamic condition across a jump, since it involves both stress and velocity.

We turn to a relation, which governs the kinematics of discontinuity. It is convenient to introduce a coordinate system, Lt, with L being a distance to the surface along the normal and t being time (Figure 1-17). A curve given by the equation, $dt/dL = 1/c$, represents the jump motion along the normal according to (1.67). A solid line, PQ, is a small element of the curve. Next, introduce a continuous function, T, with the first derivatives experiencing a jump across the curve, and write down the following identity:

$$T(Q)-T(R)+T(R)-T(P) = T(Q)-T(S)+T(S)-T(P) \quad . \tag{1.74}$$

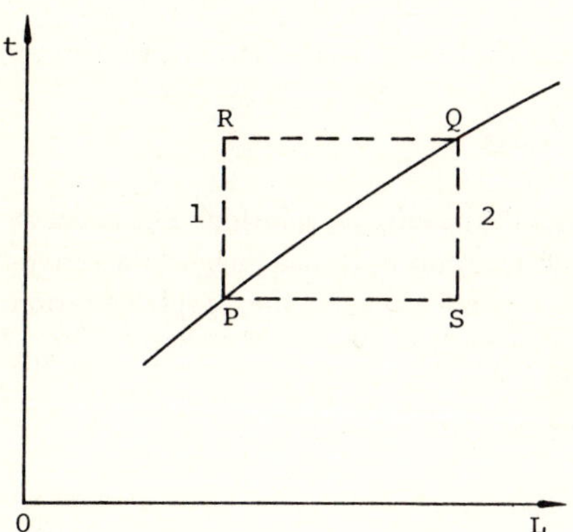

Fig. 1-17 Motion in the Lt – coordinates.

According to the mean value theorem

$$T(Q)-T(R) \simeq (\partial T/\partial L)dL \quad ,$$

where $dL \simeq RQ$ and $\partial T/\partial L$ is evaluated for an intermediate point. Similarly

$$T(R)-T(P) \simeq (\partial T/\partial t)dt$$

$$T(Q)-T(S) \simeq (\partial T/\partial t)dt$$

$$T(S)-T(P) \simeq (\partial T/\partial L)dL \quad .$$

Substituting these relations in (1.74), dividing by dt, and passing to the limit $Q \rightarrow P$, one gets

$$(\frac{\partial T}{\partial t})_2 + c(\frac{\partial T}{\partial L})_2 = (\frac{\partial T}{\partial t})_1 + c(\frac{\partial T}{\partial L})_1 \quad ,$$

where subscripts 1 and 2 denote the values immediately behind and in front of the surface. Now invoking

$$\frac{\partial T}{\partial L} = n_i \frac{\partial T}{\partial x_i} \quad ,$$

one arrives at the Hadamard kinematic condition,

$$\left[\frac{\partial T}{\partial t}\right] + c\left[\frac{\partial T}{\partial x_i}\right] n_i = 0 \quad .$$

In particular, for $T = u_j$, this expression yields the relation for derivatives of the displacement across a jump surface, $S(t)$,

$$\left[\frac{\partial u_j}{\partial t}\right] + c\left[\frac{\partial u_j}{\partial x_i}\right]n_i = [v_j] + c[u_{j,i}]n_i = 0 \tag{1.75}$$

which is the compatibility equation for shock waves. It is surprising that an expression can be deduced, which relates the jumps of time – and space – derivatives of the displacement, regardless of the type of solid, as (1.75) does.

1.13 Characteristics and Hyperbolicity

An equation may often be appreciated without knowing its solution. The concept of characteristics is helpful in this sense. Furthermore, it also suggests a convenient method of solution. To illustrate the concept consider the simple equation

$$\ddot{u} - c^2 u_{,xx} = 0 \ , \quad c = \text{const} \tag{1.76}$$

which describes, for example, vibrations of an elastic string.

It can be rewritten as

$$\left(\frac{\partial}{\partial x} + \frac{1}{c}\frac{\partial}{\partial t}\right)\left(\frac{\partial}{\partial x} - \frac{1}{c}\frac{\partial}{\partial t}\right)u = 0 \quad . \tag{1.77}$$

To integrate (1.77) we assume the existence of new independent variables, $\theta(x,t)$ and $\Omega(x,t)$, for which

$$\frac{\partial}{\partial \Omega} = \frac{\partial}{\partial x} + \frac{1}{c}\frac{\partial}{\partial t} \ , \quad \frac{\partial}{\partial \theta} = \frac{\partial}{\partial x} - \frac{1}{c}\frac{\partial}{\partial t} \quad . \tag{1.78}$$

Equation (1.77) then becomes

$$\frac{\partial^2 u}{\partial \theta \partial \Omega} = 0 \quad ,$$

with the solution

$$u = f(\theta) + g(\Omega) \quad . \tag{1.79}$$

It remains to find the functions, $\theta(x,t)$ and $\Omega(x,t)$. To this end, we set

$$\frac{\partial}{\partial x} = \frac{\partial \Omega}{\partial x}\frac{\partial}{\partial \Omega} + \frac{\partial \theta}{\partial x}\frac{\partial}{\partial \theta} \tag{1.80}$$

$$\frac{\partial}{\partial t} = \frac{\partial \Omega}{\partial t}\frac{\partial}{\partial \Omega} + \frac{\partial \theta}{\partial t}\frac{\partial}{\partial \theta} \quad . \tag{1.81}$$

Then the first of (1.78) provides

$$\frac{\partial}{\partial \Omega} = \left(\frac{\partial \Omega}{\partial x} + \frac{1}{c}\frac{\partial \Omega}{\partial t}\right)\frac{\partial}{\partial \Omega} + \left(\frac{\partial \theta}{\partial x} + \frac{1}{c}\frac{\partial \theta}{\partial t}\right)\frac{\partial}{\partial \theta} \quad ,$$

which requires

$$\frac{\partial \Omega}{\partial x} + \frac{1}{c}\frac{\partial \Omega}{\partial t} = 1, \qquad \frac{\partial \theta}{\partial x} + \frac{1}{c}\frac{\partial \theta}{\partial t} = 0 \quad . \tag{1.82}$$

Similarly, the second of (1.78) yields

$$\frac{\partial \Omega}{\partial x}-\frac{1}{c}\frac{\partial \Omega}{\partial t}=0, \qquad \frac{\partial \theta}{\partial x}-\frac{1}{c}\frac{\partial \theta}{\partial t}=1 \quad . \tag{1.83}$$

It is easily seen that the solutions to (1.82) and (1.83) are

$$\theta=\frac{1}{2}(x-ct) \, , \qquad \Omega=\frac{1}{2}(x+ct) \quad . \tag{1.84}$$

According to (1.79) the solution to (1.76) is

$$u=f(x-ct)+g(x+ct) \quad , \tag{1.85}$$

with f and g arbitrary functions.

The value of f does not change provided that positions x_1, x_2 and times t_1, t_2, are related by $x_2-x_1=c(t_2-t_1)$, which means that the wave moves from left to right with the speed c. Similarly, the displacement $g(x+ct)$ moves in the opposite direction with the same speed.

Equation (1.85) shows that at $t=0$ the two disturbances are given by the curves $f(x)$ and $g(x)$. We may therefore construct the string position for $t>0$ by moving properly these curves along the x-axis with the speed c and summing them up.

The independent variables, $x-ct=S_1$ and $x+ct=S_2$, which form a net in the plane (x,ct), are known as characteristics. In view of (1.82) and (1.83) the condition imposed on, say, the θ-characteristic is

$$c^2(\theta_{,x})^2-(\theta_{,t})^2=0 \quad . \tag{1.86}$$

Another view on the characteristics stems from the following procedure. Equation (1.76) with the help of (1.80) and (1.81) can be rewritten as

$$\begin{aligned}(c^2\theta^2_{,x}-\theta^2_{,t})u_{,\theta\theta}&+(c^2\Omega_{,x}\theta_{x}-\Omega_{,t}\theta_{,t})u_{,\Omega\theta}\\ &+(c^2\Omega^2_{,x}-\Omega^2_{,t})u_{,\Omega\Omega}+(c^2\theta_{,xx}-\theta_{,tt})u_{,\theta}\\ &+(c^2\Omega_{,xx}-\Omega_{,tt})u_{,\Omega}=0 \quad .\end{aligned} \tag{1.87}$$

Take points, $\theta=S_1\pm\varepsilon$, $\Omega=S_2$, then write for them the above equation, subtract and set $\varepsilon\to 0$ to obtain

$$\{c^2(\theta_{,x})^2-(\theta_{,t})^2\}\lim_{\varepsilon\to 0} u_{,\theta\theta}\Big|_{S_1-\varepsilon}^{S_1+\varepsilon}=0 \quad , \tag{1.88}$$

which reduces to (1.86) provided that $u_{,\theta\theta}$ is discontinuous. Recalling the discussion of Section 1.11, it follows that a characteristic may represent a wave front.

A similar procedure, applied to a more general equation

$$au_{,xx}+bu_{,xt}+du_{,tt}=0 \tag{1.89}$$

with constant a, b and d, shows

$$a(\theta_{,x})^2+b\theta_{,x}\theta_{,t}+d(\theta_{,t})^2=0 \quad ,$$

if $u_{,\theta\theta}$ is discontinuous. Considering this equation as a quadratic one in the ratio $\theta_{,x}/\theta_{,t} = \alpha$, we get

$$\alpha_{1,2} = \frac{-b\pm(b^2-4ad)^{1/2}}{2a} \quad ,$$

which shows that characteristics are real provided $b^2-4ad > 0$. Accordingly, (1.89) is said to be

$$\begin{aligned} &\text{hyperbolic,} \quad \text{if} \quad b^2 > 4ad \\ &\text{parabolic,} \quad \text{if} \quad b^2 = 4ad \\ &\text{elliptic,} \quad \text{if} \quad b^2 < 4ad \quad . \end{aligned}$$

Hence, the process of wave propagation may be described by equations of the hyperbolic type. Since the type of boundary and initial conditions, needed to insure the uniqueness of the solution, depends on the type of equation, the above classification is of importance. For example, elliptic equations, such as Laplace's equation, require the specification of the function or its normal derivative on the entire boundary of the domain. Unlike this, for the wave equation, which is hyperbolic, the function and its time derivative may be specified over a part of the boundary only.

The next reason, the above classification is essential for, is the capability of predicting a kind of functions governed by the equation. For example, the heat equation, which is parabolic, implies rapid smoothening of initial discontinuities. On the other hand, the wave equation "supports" shock propagation.

It has been noted in the foregoing that a characteristic may form the boundary between disturbed and undisturbed regions. It is therefore closely related to Huygens' principle, which suggests the following procedure for constructing a progressive front: treat every point of a front given at time t as a secondary source with its own wave surface and construct the front at $t + \Delta t$ as the envelope of secondary wave surfaces.

1.14 Energy Flux

The total energy (which is the sum of the kinetic and potential energies) contained within a wave front, progressing in a lossless medium is given by

$$E_{tot} = K_{tot} + W_{tot} = \frac{1}{2}\int_{V_1} (\rho\dot{u}_i\dot{u}_i + c_{ijk\ell}u_{i,j}u_{k,\ell})dv \quad , \tag{1.90}$$

where V_1 denotes the disturbed region (see Figure 1-15). Here use has been made of (1.31) and results of Section 1.10. This provides

$$\frac{\partial E}{\partial t} = \int_{V_1} (\rho\dot{u}_i\ddot{u}_i + c_{ijk\ell}\dot{u}_{i,j}u_{k,\ell})dv \quad , \tag{1.91}$$

which should represent the energy crossing a wave surface in unit time. On introducing the energy-flux vector, P_j, we set

$$\frac{\partial E}{\partial t} = -\int_S P_j n_j ds \quad ,$$

with S denoting a wave surface.

Equations (1.15) and (1.31) yield for the absence of body forces, $f_i = 0$,

$$\rho \ddot{u}_i = c_{ijk\ell} u_{k,\ell j}$$

and (1.91) may then be rewritten as

$$\int_{V_1} (\dot{u}_i c_{ijk\ell} u_{k,\ell j} + c_{ij\ell} \dot{u}_{i,j} u_{k,\ell}) dv$$

$$= \int_{V_1} (c_{ijk\ell} u_{k,\ell} \dot{u}_i),_j dv$$

$$= \int_S c_{ijk\ell} u_{k,\ell} \dot{u}_i n_j ds = -\int_S P_j n_j ds \quad .$$

This provides

$$P_j = -c_{ijk\ell} u_{k,\ell} \dot{u}_i \quad .$$

In view of (1.31), P_j, known as the Poynting vector, can be rewritten as

$$P_j = -\sigma_{ij} \dot{u}_i \quad . \tag{1.92}$$

This vector describes the direction and rate of the energy transport. It should be pointed out that, in general, $\boldsymbol{P}$ is not normal to the wave surface, S.

1.15 Basic Concepts of Waves. Debye's Frequency

A wave equation (1.76), which has been often dealt with in previous sections, provides a convenient model for illustrating basic concepts of wave motion. Its solutions can be written in different ways, the fact reflecting a variety of possible physical situations. One of them, known as a running wave, is as follows:

$$u = f(x \pm ct) \quad , \tag{1.93}$$

in accord with (1.85). The explicit form of this function may be found from additional information available. The above expression describes a disturbance propagating with velocity c and maintaining its shape. In fact, f may be represented by a curve, as shown in Figure 1-18, when time t is given. This function, as well as its argument, preserve their values provided the increments in time, Δt, and in distance, Δx, are related by

$$|\Delta x|/\Delta t = c$$

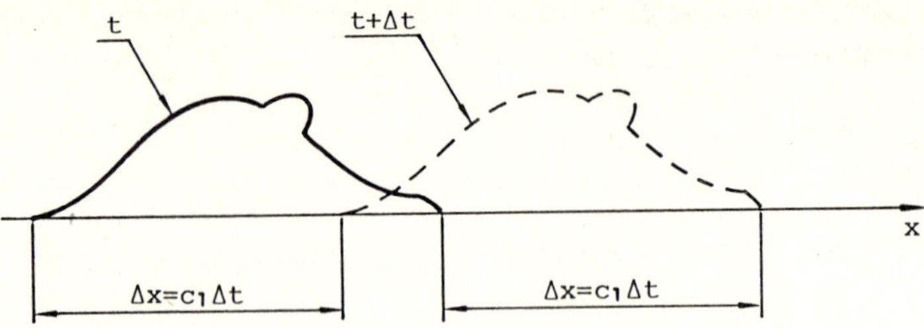

Fig. 1-18 Wave motion along a string.

as has been described in Section 1.13

A particular case of (1.93) is given by

$$u = u_o e^{(i\omega/c)(x \pm ct)} = u_o e^{i(kx \pm \omega t)}$$

known as a free harmonic wave with the circular frequency ω and the wave number $k = \omega/c$. The quantity $(kx \pm \omega t)$ is referred to as the phase, whereas $u_o e^{ikx}$ as the complex amplitude. Of course, only real or imaginary part of the complex solutions is physically meaningful and measurable. In the framework of linear operations the use of complex functions proved convenient. However, to compute the energy or power one has, first, to pass to real values. Also, it is common to use combinations of complex values. The ratio of the response to harmonic excitation is known as the admittance functions, while the ratio of stress to velocity as the impedance.

Additional characteristics used to describe harmonic disturbances are

$$\begin{aligned} f &= \omega/(2\pi), && = \text{cyclic frequency} \\ \lambda &= 2\pi/k, && = \text{wavelength} \\ T &= 1/f, && = \text{period} \quad . \end{aligned}$$

A special case of harmonic disturbance is known as a standing wave. Consider a sum of leftward and rightward running disturbances with the equal amplitudes,

$$u(x,t) = \frac{u_o}{2}\cos(kx + \omega t) + \frac{u_o}{2}\cos(kx - \omega t) = u_o \cos kx \cos \omega t \quad .$$

Clearly, the points x_n, which are given by

$$\cos kx_n = 0 \ , \quad x_n = n\pi/(2k) \ , \quad n = 1, 3, 5...$$

have zero displacement. The presence of such points, called nodes, shows that the interference of running disturbances may generate stationary vibrations of a system.

A wave with an arbitrary time-dependence may be represented by the Fourier series or the Fourier integral. As is well known, a periodic function with period

$T = 2\pi/\omega_o$ expands into a series

$$u(t,x) = \frac{1}{\sqrt{2\pi}} \sum_{n=-\infty}^{\infty} u_n(x) e^{-in\omega_o t} \quad ,$$

where the $u_n(x)$ constitute the discrete amplitude spectrum at a point x and are given by

$$u_n(x) = \frac{\omega_o}{\sqrt{2\pi}} \int_o^T u(t,x) e^{in\omega_o t} dt \quad .$$

An aperiodic function under some regularity conditions may be represented as

$$u(t,x) = \frac{1}{\sqrt{2\pi}} \int_{-\infty}^{\infty} u(\omega,x) e^{-i\omega t} d\omega \quad , \tag{1.94}$$

with $u(\omega, x)$ denoting the spectral density given by

$$u(\omega,x) = \frac{1}{\sqrt{2\pi}} \int_{-\infty}^{\infty} u(t,x) e^{i\omega t} dt \quad . \tag{1.95}$$

It can be shown that an elementary wave, either $u_n(x)e^{-i\omega t}$ or $u(\omega, x)e^{-i\omega t}d\omega$, satisfies the wave equation (1.76), provided the function $u(t, x)$ does (see comments on Problem 1-20). That is why harmonic waves play a central role in linear theory.

The following example explains the concept of wave dispersion. A transverse disturbance, w, propagating along a beam subjected to small deformations is governed by the equation

$$\frac{\partial^4 w}{\partial x^4} + \frac{1}{a_1^2} \frac{\partial^2 w}{\partial t^2} = 0 \ , \quad a_1^2 = EI/(\rho A) \quad ,$$

with ρ, A, E, and I denoting the mass density, the cross-sectional area, Young's modulus, and the inertia moment, respectively. Substituting

$$w \simeq e^{(-i\omega t + ikx)}$$

yields

$$k^4 = \omega^2/a_1^2 \quad .$$

On the other hand, it follows from the discussion of running waves that their velocity is given by

$$c_1 = \omega/k \quad .$$

Thus

$$c_1 = (\omega a_1)^{1/2} \quad . \tag{1.96}$$

Note, that for the string, considered earlier, the velocity is

$$c = \omega/k = \text{const} \quad . \tag{1.97}$$

Equations of the type (1.96) are referred to as the dispersion relations. Equation (1.96) indicates that the waves in the string are qualitatively different from those in the beam. In fact, in the latter case, the velocity depends on frequency. This phenomenon is known as the dispersion. It may be caused by the geometry of a waveguide as well as by physical properties of a material.

A continuum, as a mathematical abstraction has an infinite number of degrees of freedom, which puts no upper limit on frequency. In fact, the frequency interval is usually taken infinite, $0 \leq \omega < \infty$. On the other hand, it is intuitively clear that no real solid medium can support vibrations of infinitely large frequency. By taking into account the discrete atomic structure of real solids, Debye has shown that the upper cut-off frequency, $\omega_D > \omega$, is given by

$$\omega_D = 2\pi c(\frac{3}{4\pi V_a})^{1/3}$$

where c is the wave velocity and V_a the atom volume. Debye's frequency, which assumes an ordered microstructure, is usually very high, $\omega_D \simeq 10^{12} - 10^{14} 1/sec$.

1.16 Group and Signal Velocities

The dependence of the wave speed on frequency, $c(\omega)$, typical of solids with a microstructure as well as of waveguides, has a variety of consequencies. In particular, it makes questionable the very concept of the wave speed.

If the medium is nondispersive, then any wave propagates with the same velocity. This fact stems from the superposition principle and the possibility of viewing a wave as a composition of harmonic components having the same speed, c. Since this concept arises from the condition of equal phases for t and $t + \Delta t$ at x and $x + \Delta x$, respectively,

$$kx \mp \omega t = \frac{\omega}{c}(x \pm \Delta x) \mp \omega(t + \Delta t) \to c = \frac{\omega}{k} \quad ,$$

c is also called a phase velocity.

A different situation prevails, when a medium is dispersive. In this case the superposition principle still holds, but each harmonic component propagates with its own velocity, according to the dispersion relation

$$c = c(\omega, k) \quad .$$

It may seem that the notion of the wave velocity breaks down. Nevertheless, under certain conditions it applies to dispersive wave too.

We may arrive at a proper concept by starting with a superposition of two running disturbances,

$$u = u_o[\cos(k_1 x - \omega_1 t) + \cos(k_2 x - \omega_2 t)] \quad . \tag{1.98}$$

If $\omega_1 = \omega_2 = \omega$ and, hence, $k_1 = k_2 = k$, the speed is given by

$$c(\omega, k) = \omega/k \quad .$$

Now consider the case when the frequencies are slightly different

$$\omega_2 - \omega_1 = \Delta\omega \ , \qquad k_2 - k_1 = \Delta k \quad ,$$

where Δ designates a small value. Equation (1.98) can then be rewritten as

$$u = 2u_o \cos\{\frac{1}{2}[(k_2-k_1)x - (\omega_2-\omega_1)t]\} \cos\{\frac{1}{2}[(k_1+k_2)x - (\omega_1+\omega_2)t]\}$$
$$= u_o \cos(\frac{1}{2}\Delta kx - \frac{1}{2}\Delta\omega t) \cos(\tilde{k}x - \tilde{\omega}t) \quad ,$$

with

$$\tilde{k} = \frac{1}{2}(k_1 + k_2) \, , \qquad \tilde{\omega} = \frac{1}{2}(\omega_1 + \omega_2) \quad .$$

One observes from the right-hand side of this expression that the disturbance consists of two terms: the first, with low frequency, $\Delta\omega/2$, which is referred to as the modulation, and the second one of higher frequency, $\tilde{\omega}$, referred to as the carrier. The speed of modulation wave, c^m, is given by

$$c^m = \Delta\omega/\Delta k \quad ,$$

whereas that of the carrier by

$$c^c = \tilde{\omega}/\tilde{k} \quad .$$

There exists no unique relation between these two speeds. From Figure 1-19

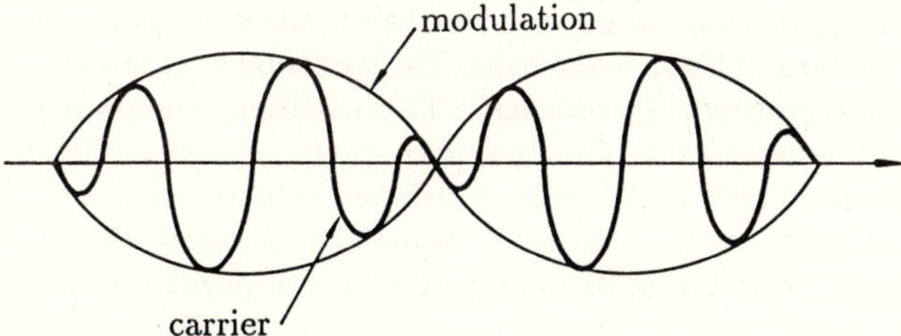

Fig. 1-19 Modulation and carrier.

follows that the overall motion of the wave pair is characterized by the velocity of the modulation term, c^m. Referring to this value as the group velocity, c_g, one defines it as the limit of the modulation speed,

$$c_g = \lim_{\Delta k \to 0} c^m = \lim_{\Delta k \to 0} \frac{\Delta\omega}{\Delta k} = \frac{d\omega}{dk} \quad . \tag{1.99}$$

Thus, we have established the meaningful concept of the speed of a disturbance consisting of two slightly different frequencies. Now we consider a superposition (wave packet),

$$u = u_1 e^{-i\omega_1 t + ik_1 x} + u_2 e^{-i\omega_2 t + ik_2 x} + \ldots = \sum_n u_n e^{-i\omega_n t + ik_n x} \quad ,$$

where the frequencies and wave numbers are to satisfy the conditions

$$|k_j - k_i| \ll k_1 \, , \quad (\omega_j - \omega_i) \ll \omega_1 \quad .$$

Then the superposition can be written as

$$u = e^{(-i\omega_1 t + ik_1 x)}\{u_1 + u_2 e^{-i(\omega_2-\omega_1)t+i(k_2-k_1)x} + u_3 e^{-i(\omega_3-\omega_1)t+i(k_3-k_1)x} + \ldots\}$$
$$= e^{(-i\omega_1 t + ik_1 x)} \sum_n u_n e^{-i(\omega_n-\omega_1)t+i(k_n-k_1)x} \quad .$$

On taking into account

$$\frac{\omega_2-\omega_1}{k_2-k_1} \simeq \frac{\omega_3-\omega_1}{k_3-k_1} \simeq \quad \ldots \quad \simeq \frac{d\omega}{dk} = c_g$$

and neglecting the higher order terms one gets

$$u = e^{(-i\omega_1 t + ik_1 x)}\{u_1 + u_2 e^{i(k_2-k_1)(x-c_g t)} + \ldots\}$$
$$= e^{(-i\omega_1 t + ik_1 x)} \sum_n u_n e^{-i(k_n-k_1)(x-c_g t)} \quad .$$

Thus, the disturbance is again represented as a carrier propagating with the speed $c_1 = \omega_1/k_1$, and the modulation wave with the group velocity, c_g. One concludes that as far as the wave contains a narrow band of frequencies, its propagation can be fairly described by the group velocity given by (1.99).

The propagation speed of a wide-band disturbance in a dispersive medium cannot be characterized by any single number. Nevertheless, the concept of group velocity does apply to analysis of this phenomenon. The spectrum of such a signal may be divided into small intervals of frequencies. This partition represents the wave as a composition of wave packets with their own group velocities. Clearly, the location of wave packets within the pulse undergoes redistribution during propagation: the faster ones travel to the front, while the slower to the rear. Thus, the pulse will spread out during its propagation in a dispersive medium, which could create no sharp wave front. Nevertheless, it can be shown that a well-defined wave front does exist provided that the limit

$$c_\infty = \lim_{\omega\to\infty} c(\omega) \tag{1.100}$$

is finite.

To this end, consider a waveguide, say, a dispersive half-space, $x \geq 0$, subjected to a displacement, $u_o(t)$ at $x = 0$. To find the progressive disturbance, $u(x,t)$, we first determine the Fourier transform of $u_o(t)$

$$u(\omega) = \frac{1}{\sqrt{2\pi}} \int_{-\infty}^{\infty} u_o(t) e^{i\omega t} dt \quad . \tag{1.101}$$

Then, the superposition of plane waves, $u(\omega)e^{i(kx-\omega t)}/\sqrt{2\pi}$, provides

$$u(x,t) = \frac{1}{\sqrt{2\pi}} \int_{-\infty}^{\infty} u(\omega) e^{i(kx-\omega t)} d\omega \quad , \tag{1.102}$$

where $k = \omega/c(\omega)$.

If $u_o(t)$ is zero for $t < 0$, then all the singularities of the integrand in (1.102) can be shown to lie on or below the real axis. Thus, for $(t - x/c_\infty) < 0$, the integration contour is located in the upper half-plane, above the singularities, and includes arc of infinite radius. This implies that the displacement, $u(x,t)$, is zero for $(t-x/c_\infty) < 0$. At the instant $t = x/c_\infty$ a point x will first learn that the disturbance has been launched at the boundary, $x = 0$. Note, that only the existence of the infinite-frequency limit, c_∞, matters, while no similar condition is imposed on intermediate frequencies. Details of a proper analysis will be given in Section 2.15, in the context of a viscoelastic pulse propagation. The limit, c_∞, which describes the propagation speed of the wave front is referred to as the signal velocity. The above considerations are closely related to the concept of causality outlined in what follows.

1.17 Causality and General Relations for Linear Systems

A "philosophical" observation that the output cannot precede the input, known as the primitive causality condition, when used properly, gives rise to remarkable general relations governing the behavior of linear damped systems.

The well-known convolution integral

$$g(t) = \int_{-\infty}^{\infty} F(\tau)u^\delta(t-\tau)d\tau = \int_{-\infty}^{\infty} F(t-\tau)u^\delta(\tau)d\tau \tag{1.103}$$

relates the input, $F(t)$, and the output, $g(t)$, via the impulse response, $u^\delta(t)$, of a linear system. If a time function equals zero for $t < 0$, it is called a causal function. Accordingly, for a causal system, $u^\delta(t-\tau) = 0$ for $t < \tau$, and (1.103) provides

$$g(t) = \int_0^t F(\tau)u^\delta(t-\tau)d\tau = \int_0^t F(t-\tau)u^\delta(\tau)d\tau \quad . \tag{1.104}$$

The impulse response, $u^\delta(t)$, may be easily related to the admittance function, $H(i\omega) = H^r(\omega) + iH^i(\omega)$, of the system. Indeed, by making use once again of the superposition principle, we get, similarly to (1.102),

$$g(t) = (2\pi)^{-1/2} \int_{-\infty}^{\infty} F(\omega)H(i\omega)e^{-i\omega t}d\omega \quad , \tag{1.105}$$

where $F(\omega)$ is the Fourier transform of the input, $F(t)$. On choosing $F(t) = \delta(t)$, and, accordingly, $F(\omega) = (2\pi)^{-1/2}$, (1.105) yields

$$u^\delta(t) = (2\pi)^{-1} \int_{-\infty}^{\infty} H(i\omega)e^{-i\omega t}d\omega \quad , \tag{1.106}$$

which is a source of many other useful relations. (In general, the above two integrals should be understood as principal values).

For example, on representing $u^\delta(t)$ as

$$u^\delta(t) = u_1^\delta(t) + u_2^\delta(t) \quad ,$$

where the first term in the right-hand side is an even function and second an odd function, and inverting (1.106), we get

$$\operatorname{Re} H(i\omega) = 2\int_0^\infty u_1^\delta(t)\cos\omega t dt$$

$$\operatorname{Im} H(i\omega) = 2\int_0^\infty u_2^\delta(t)\sin\omega t dt$$

and, similarly

$$u_1^\delta(t) = \int_0^\infty \operatorname{Re} H(i\omega)\cos\omega t d\omega/\pi$$

$$u_2^\delta(t) = \int_0^\infty \operatorname{Im} H(i\omega)\sin\omega t d\omega/\pi \quad .$$

Next, due to the causality, $u^\delta(-t) = 0$, and

$$u_1^\delta(-t) + u_2^\delta(-t) = u_1^\delta(t) - u_2^\delta(t) = 0 \, , \qquad (t > 0)$$

which implies

$$\begin{aligned} u^\delta(t) &= 2\int_0^\infty \operatorname{Re} H(i\omega)\cos\omega t d\omega/\pi \\ &= 2\int_0^\infty \operatorname{Im} H(i\omega)\sin\omega t d\omega/\pi \end{aligned} \tag{1.107}$$

for $t > 0$.

The above relation is particularly instructive, since it imposes certain analytical behavior of $\operatorname{Re} H(i\omega)$ and $\operatorname{Im} H(i\omega)$ and a dependence between them, following from the linearity and the causality of a system. To this end, assume that the admittance function, $H(i\omega)$, is bounded and analytic in the upper half-plane. As a function of a complex argument, which is not a constant, it has the singularities located in the lower half-plane. For convenience, we shall also assume that $H(i\omega)$ does not have poles on the real axis.

Consider a contour in the upper half-plane of the variable $\Omega = \omega + i\omega'$, which is pictured in Figure 1-20. Here R_1 and R_2 are the radii of the large and small semicircles, Q_1 and Q_2, respectively. Then, according to Cauchy's theorem

$$0 = \int_Q \frac{H(i\Omega)d\Omega}{\Omega - \omega} = \int_{Q_1} + \int_{-R_1}^{\omega - R_2} + \int_{Q_2} + \int_{\omega + R_2}^{R_1} \left[\frac{H(i\Omega)d\Omega}{\Omega - \omega}\right] \quad .$$

On taking the limits $R_1 \to \infty$, $R_2 \to 0$, and separating the real and imaginary parts, we get the Kramers-Kronig relations

$$\begin{aligned} \operatorname{Re} H(i\omega) &= \operatorname{Re} H(\infty) - P\int_{-\infty}^\infty \frac{\operatorname{Im} H(i\varepsilon)d\varepsilon}{(\varepsilon - \omega)\pi} \\ \operatorname{Im} H(i\omega) &= \operatorname{Im} H(\infty) + P\int_{-\infty}^\infty \frac{\operatorname{Re} H(i\varepsilon)d\varepsilon}{(\varepsilon - \omega)\pi} \quad , \end{aligned} \tag{1.108}$$

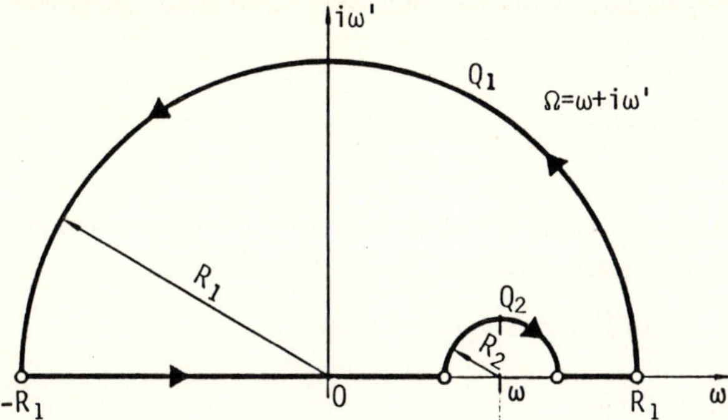

Fig. 1-20 Integration in the complex plane.

with P denoting Cauchy principal value. As this derivation shows, the expressions above rely solely on the causality, linearity and passivity of the system, without any appeal to its specific nature. The causality is related to analiticity of the admittance function in the upper half-plane, linearity gives rise to the convolution integral (1.103), and passivity (location of the poles in the lower half-plane) is due to internal energy losses and stability.

Equations (1.108) may be represented in different forms and applied to a variety of systems. In the case of a viscoelastic modulus, $G(i\omega) = H(i\omega)$, they transform to (1.47). Similar relations may also be written for the complex wave number, $k(i\omega)$, describing wave propagation in attenuative media, such as viscoelastic or inhomogeneous materials.

Problems

1-1 Denote rays passing through the middle point of faces of a cubic system by means of indices.

1-2 Write down the indices of a face plane of a cubic system.

1-3 Denote the diagonal plane of a cubic system by means of indices.

1-4 The field of displacement due to the presence of an edge dislocation (see Figure 1-6) is given by

$$u_1 = \frac{b}{2}\pi[\tan^{-1}(x_1/x_2) + \frac{1}{2(1-\nu)}x_1x_2/r^2]$$

$$u_2 = -\frac{b}{4\pi(1-\nu)}[(1-2\nu)\log r + x_2^2/r^2]$$

with $r^2 = x_1^2 + x_2^2$. Find the stresses.

1-5 The field of displacement due to the presence of a screw dislocation (see Figure 1-7) is given by

$$u_3 = \frac{b}{2\pi} \tan^{-1}(x_1/x_2) \quad ,$$

where u_3 is the displacement along AB-direction and x_1, x_2 are cartesian coordinates in a horizontal plane. Analyze the discontinuity available.

1-6 The field of velocities, v_i, in Eulerian space is given by

$$v_i = d_{ij}(t)X_j \quad .$$

Show that the displacement vector is a linear function of Lagrangian coordinates.

1-7 Show that the deformation measures introduced in Section 1.4 are reduced to (1.3) provided $(\ell - \ell_o)/\ell_o \rightarrow 0$.

1-8 Outline the procedure for obtaining the invariants of a second-rank tensor, like ε_{ij} or σ_{ij}.

1-9 Deduce Hooke's law from (1.28) and (1.29).

1-10 Derive the explicit form of Hooke's law for orthotropic media.

1-11 Outline the procedure for obtaining (1.46) from (1.45).

1-12 Derive the equations similar to (1.50) and (1.55), making use of ε_{ij}, E_i, and T as independent variables, and σ_{ij}, D_i, and S as dependent ones.

1-13 Determine the relations among the adiabatic and isothermal coefficients for the coupled medium considered in Section 1.9.

1-14 Find the Euler equations associated with Hamilton's principle.

1-15 Justify (1.66).

1-16 Write down the expression for the energy flux in the direction normal to the wave surface.

1-17 Derive (1.88) from (1.87)

1-18 Consider the time average of the Poynting vector for the case of one-dimensional standing waves and find the net flow of energy.

1-19 Deduce the expression of the wave velocity for an elastic string by means of direct analysis of the forces involved. Does the result depend on the waveform?

1-20 Show that the elementary component of the Fourier representation of a function $f(r,t)$ satisfies the wave equation provided the function does.

1-21 A wave consists of two frequencies, ω_1 and ω_2, only. Is it expandable in a Fourier series if $\omega_1 \simeq \omega_2$?

1-22 Is (1.76) invariant with respect to the transformation $x = (x' + at)$ with $a = \text{const}$? Explain the result.

References and Additional Reading

Section 1.2

Kosevich (1979), Van-Vlack (1980).

Section 1.3 – 1.7

Eringen and Suhubi (1974, 1975), Green and Zerna (1968), Sokolnikoff (1956). See Willis (1981) for the microstructure description and Seth (1966) for the analysis of the generalized strain measure.

Section 1.8

Bert (1973), Christensen (1971), Pipkin (1972).

Section 1.9

Auld (1973), Dieulesaint and Royer (1980), and Tucker and Rampton (1972) provide a complete investigation of piezoelectricity.

Section 1.10

Bedford (1985) provides a comprehensive treatment of Hamilton's principle, Sokolnikoff (1956).

Section 1.11 – 1.13

Miklowitz (1978). Nowacki (1978) investigates non-linear waves in solids.

Section 1.14

Dieulesaint and Royer (1980), Miklowitz (1978).

Section 1.15

Morse and Ingard (1968). See Kubo and Nagamiya (1969) for the concept of Debye's frequency.

Section 1.16

Morse and Ingard (1968)

Section 1.17

Nussenzveig (1972) presents an exaustive analysis of the Kramers-Kronig relations.

Chapter II

Bulk Waves in Isotropic Media

Because of its abstract nature the concept of an unbounded medium may provide a unique possibility of investigating basic features of wave motions in an exact and, at the same time, relatively simple way.

In the first part of this chapter we establish the fundamental fact of existence of two different modes of elastic waves in an isotropic medium and deal with their mathematical representations. The subject may therefore appear somewhat formal. Nevertheless, general solutions derived are then of systematic use in the second part, which is concerned with elastic and viscoelastic radiation from a variety of sources.

The preference was given to radiation from cavities, rigid inclusions, and microdefects. Besides their obvious relevance as physical models, these sources possess simple solutions, which display such phenomena as radiation damping, resonance and wavefront speed in an easily interpretable way. Both, steady-state and transient waves, are considered. These solutions also facilitate the understanding of wave-obstacle interactions considered in Chapter 5. Green's dynamic tensors are given because of their major role in Elastodynamics. To avoid calculational "over-loading", some of the derivations were left to exercises.

This chapter is completed by a brief consideration of a simple case of shock waves. This mainly purports to stress the limited nature of the linear theory, which is, nevertheless, of extreme usefulness in many practical cases.

2.1 Modes and Velocities of Elastic Waves

It has been shown in Section 1.5 that the equations of motion in a continuum are given by

$$\rho \ddot{u}_i = \sigma_{ij,j} \quad ,$$

which hold provided the body forces are neglected. Let us adapt it for the particular case of isotropic elastic media. Substituting (1.42) for the stress, one has, say, for $i = 1$

$$\rho\ddot{u}_1 = \frac{\partial}{\partial x_1}(\lambda\tilde{\theta} + 2\mu\varepsilon_{11}) + \frac{\partial}{\partial x_2}(\mu\varepsilon_{12}) + \frac{\partial}{\partial x_3}(\mu\varepsilon_{13}) \quad .$$

By invoking (1.9) for the strain the relation above can be transformed to the following equation:

$$\rho\ddot{u}_1 = (\lambda+\mu)\partial\tilde{\theta}/\partial x_1 + \mu\nabla^2 u_1 \quad , \tag{2.1}$$

where ∇^2 denotes $\partial^2/\partial x_1^2 + \partial^2/\partial x_2^2 + \partial^2/\partial x_3^2$ and $\tilde{\theta} = \varepsilon_{ii}$. Similarly, for the displacements u_2 and u_3 one gets

$$\begin{aligned} \rho\ddot{u}_2 &= (\lambda+\mu)\partial\tilde{\theta}/\partial x_2 + \mu\nabla^2 u_2 \\ \rho\ddot{u}_3 &= (\lambda+\mu)\partial\tilde{\theta}/\partial x_3 + \mu\nabla^2 u_3 \quad . \end{aligned} \tag{2.2}$$

Equations (2.1) and (2.2) are known as the dynamic equations of displacements for isotropic linear elastic media (the body forces are neglected) and can be conveniently written as follows:

$$\rho\ddot{u}_i = (\lambda+\mu)\tilde{\theta}_{,i} + \mu\nabla^2 u_i \quad . \tag{2.3}$$

It is useful to transform them to standard equations well-known in mathematical physics. To this end, one differentiates (2.3) with respect to x_i and sums to get the so-called wave equation for dilatation $\tilde{\theta}$

$$\partial^2\tilde{\theta}/\partial t^2 = c_\alpha^2\nabla^2\tilde{\theta} \quad , \tag{2.4}$$

where $c_\alpha = [(\lambda+\mu)/\rho]^{1/2}$ is the wave speed.

Along with the change in volume, elastic strains consist of shear. This second type of deformation is closely related to the so-called rotations, ω_i, about the axis x_i. In terms of displacements the rotations are given according to (1.12) by

$$\omega_i = -1/2 e_{ijk}u_{k,j} \quad . \tag{2.5}$$

Let us deduce the governing equation, say, for ω_1. To this end, one differentiates the first of (2.2) with respect to x_3, the second with respect to x_2, and subtracts to obtain

$$\rho\frac{\partial^2}{\partial t^2}(u_{3,2}-u_{2,3}) = \mu\nabla^2(u_{3,2}-u_{2,3}) \quad ,$$

which in view of (2.5) yields

$$\partial^2\omega_1/\partial t^2 = c_\beta^2\nabla^2\omega_1 \tag{2.6}$$

with $c_\beta^2 = \mu/\rho$. Similar relations can be derived, of course, for rotations about the axes x_2 and x_3. Other than the different notations appearing in (2.4) and (2.6), their structure is identical.

We, thus, conclude that two types of bulk waves exist in isotropic elastic media. One of them is associated with the volume changes, whereas the other one with shear motions. The velocity ratio is given by

$$\chi_o = c_\alpha/c_\beta = [(\lambda+2\mu)/\mu]^{1/2} = [(2-2\nu)/(1-2\nu)]^{1/2} \tag{2.7}$$

and thus, depends solely on Poisson's ratio. The waves propagating with velocity c_α are called the dilatational or primary waves (P-waves), while those with velocity c_β are called the shear or secondary waves (S-waves). P-waves are always faster than S-waves. It can be shown via (2.7) that the following inequality holds:

$$c_\alpha > c_\beta\sqrt{2} \quad .$$

Consider the propagation of a plane wave along the x_1-axis, setting

$$u_i = f(x_1 - ct) = f(\tau_1) \quad , \tag{2.8}$$

where c is the wave velocity. Then (2.1) and (2.2) yield

$$\begin{aligned} \rho c^2 u_1'' &= (\lambda + 2\mu)u_1'' \\ \rho c^2 u_2'' &= \mu u_2'' \\ \rho c^2 u_3'' &= \mu u_3'' \quad , \end{aligned} \tag{2.9}$$

where prime denotes differentiation with respect to τ_1. These expressions can only be satisfied in one of the two following ways: either

$$\begin{aligned} c^2 &= (\lambda + 2\mu)/\rho = c_\alpha^2 \\ u_2 &= u_3 = 0 \end{aligned} \tag{2.10}$$

or

$$\begin{aligned} c^2 &= \mu/\rho = c_\beta^2 \\ u_1 &= 0 \quad . \end{aligned} \tag{2.11}$$

Clearly, in the first case we have longitudinal waves for which polarization and propagation direction coincide. Hence, such disturbances propagate with velocity of P-waves, c_α, (like the wave of dilatation given by (2.4)). In the second case the particle displacement and the direction of propagation are mutually normal, thus constituting shear waves. The waves of rotation, given by (2.6), have the same velocity c_β. Figure 2-1 provides the illustration of polarization and propagation for P- as well as S-waves. There is no coupling between modes, which thus represent independent motions.

One should note that, as far as elastic properties are concerned, S-wave velocity is governed solely by the shear modulus of the medium, whereas the velocity of P-wave is influenced by both, the bulk and shear moduli. The reason may be appreciated by considering a small cube of material in the path of a plane longitudinal wave propagating, say, in the direction of the x_1-axis. While the x_1-dimension of the cube will alter during the wave passage, its normal cross-section will remain unchanged. There is, therefore, an alteration in both the shape and volume of any small element during propagation of the P-wave. Typical values of ρ, c_α, and c_β are given in Table 2.1.

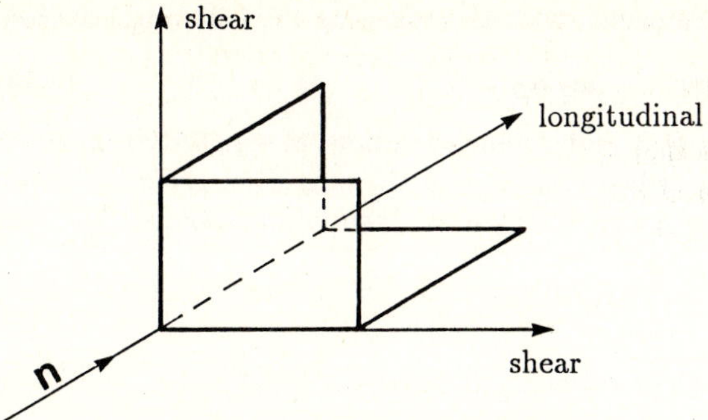

Fig. 2-1 Longitudinal and shear waves in solids.

Table 2.1 Typical values of relevant parameters

Material	ρ (kg/m^3)	c_α (m/s)	c_β (m/s)
water	1000	1480	
air	1.2	340	
steel	7800	5900	3200
aluminum	2700	6300	3100
rubber	930	1040	27
rock	2690	6000	3500

Thus, it has been shown that in the isotropic medium, the propagation direction and polarization are either parallel (P-mode) or mutually normal (S-mode). Such modes are called pure ones.* More general types of modes are typical of anisotropic media and will be investigated in Chapter 3.

The presented treatment has exploited a somewhat intuitive approach. A more powerful and rather formal analysis is given in the next section.

2.2 Vector Representation and Wave Potentials

Equation (2.3) can be rewritten in the vector way as

$$\rho \boldsymbol{u}_{,tt} = (\lambda + \mu)\,\mathbf{grad}\ \mathrm{div}\,\boldsymbol{u} + \mu \nabla^2 \boldsymbol{u} \quad . \tag{2.12}$$

Because of the identity

$$\nabla^2 \boldsymbol{u} = \mathbf{grad}\ \mathrm{div}\,\boldsymbol{u} - \mathbf{curl}\ \mathbf{curl}\,\boldsymbol{u} \tag{2.13}$$

the above equation yields

$$\rho \boldsymbol{u}_{,tt} = (\lambda + 2\mu)\,\mathbf{grad}\ \mathrm{div}\,\boldsymbol{u} - \mu\,\mathbf{curl}\ \mathbf{curl}\,\boldsymbol{u} \quad . \tag{2.14}$$

* Some authors use another definition for pure modes, which is based on energy considerations.

Let us now define two vectors u_α and u_β possessing the following features:

$$\mathbf{curl}\, u_\alpha = 0 \qquad \text{and} \qquad \operatorname{div} u_\beta = 0 \quad . \tag{2.15}$$

It is shown in vector analysis that under some conditions of regularity any vector can be represented as a sum of two vectors with the properties given by (2.15). Thus, the displacement u can be represented as a sum of dilatational (u_α) and rotational (u_β) vectors

$$u = u_\alpha + u_\beta \quad .$$

Inserting this relation into (2.14) one gets in view of (2.15)

$$\rho u_{\alpha,tt} = (\lambda + 2\mu)\, \mathbf{grad}\, \operatorname{div} u_\alpha \tag{2.16}$$

$$\rho u_{\beta,tt} = -\mu\, \mathbf{curl}\, \mathbf{curl}\, u_\beta \quad . \tag{2.17}$$

Because of (2.13) and (2.15) the following relations hold:

$$\mathbf{grad}\, \operatorname{div} u_\alpha = \nabla^2 u_\alpha \qquad \text{and} \qquad \mathbf{curl}\, \mathbf{curl}\, u_\beta = -\nabla^2 u_\beta \quad . \tag{2.18}$$

Hence, (2.16) and (2.17) provide

$$u_{\alpha,tt} - c_\alpha^2 \nabla^2 u_\alpha = 0 \tag{2.19}$$

and

$$u_{\beta,tt} - c_\beta^2 \nabla^2 u_\beta = 0 \quad , \tag{2.20}$$

where the abbreviations given by (2.10) and (2.11) are employed. Equations (2.8), derived in the previous section, are, of course, particular one-dimensional cases of the above results, which constitute the vector wave equations for the vectors u_α and u_β. In addition, comparison between (2.9) and (2.19)-(2.20) indicates the physical sense of these vectors. A longitudinal component of displacement is represented by u_α, whereas a transverse one by u_β.

To arrive at a more convenient form of governing equations one introduces the wave potentials, a scalar function Π and a vector function $\boldsymbol{F}$, by the relations

$$u_\alpha = \mathbf{grad}\, \Pi, \qquad \text{and} \qquad u_\beta = \mathbf{curl}\, \boldsymbol{F} \quad . \tag{2.21}$$

The operations involved in (2.21) can be conveniently expressed via operator, ∇, given by

$$\nabla = i_j \partial / \partial x_j \quad ,$$

where i_j are the unit vectors of cartesian reference frame. Then

$$u_\alpha = \nabla \Pi \,, \qquad \text{and} \qquad u_\beta = \nabla \times \boldsymbol{F} \quad .$$

Taking into account that

$$\operatorname{div}\, \mathbf{grad} = \nabla^2$$

and substituting the first of (2.21) into (2.16) one gets

$$\mathbf{grad}\,(\Pi,_{tt} - c_\alpha^2 \nabla^2 \Pi) = 0 \quad ,$$

which after integration with respect to coordinates provides

$$\Pi,_{tt} - c_\alpha^2 \nabla^2 \Pi = 0 \quad . \tag{2.22}$$

Similarly, the second of (2.21), and (2.17) yield

$$\boldsymbol{F},_{tt} - c_\beta^2 \nabla^2 \boldsymbol{F} = 0 \quad . \tag{2.23}$$

It should be noted that an additive solution has been lost while performing spatial integration. However, it does not represent any wave motion and, thus, has no effect on the phenomenon at hand.

The deduced equations (2.22) and (2.23) involve one scalar function and one vector function, unlike two vector functions appearing in (2.19) and (2.20), which is indisputable advantage.

Consider now a particularly simple case. Assume that the displacements are given by

$$u_1 = u_2 = 0 \,, \qquad u_3 = u_3(x_1, x_2, t) \quad . \tag{2.24}$$

The associated stresses are found from Hooke's law as follows:

$$\sigma_{13} = \mu u_{3,1} \qquad \sigma_{23} = \mu u_{3,2}$$

$$\sigma_{11} = \sigma_{22} = \sigma_{33} = \sigma_{12} = 0 \quad .$$

Then (2.21) – (2.23) reduce to the single wave equation

$$\mu \nabla^2 u_3(x_1, x_2, t) = \rho \ddot{u}_3 \quad , \tag{2.25}$$

with

$$\nabla^2 = \partial^2/\partial x_1^2 + \partial^2/\partial x_2^2 \quad .$$

Similar results can be obtained in cylindrical coordinates r, θ, z, with axis z taken along the x_3-direction. Clearly, (2.25) defines a shear horizontal or the so-called SH-wave to be further considered in Section 2.4. Our interest in this case of elastic waves lies in its simplicity and analogy with conventional sound waves, which are also governed by a single scalar wave equation. However, this analogy is solely a mathematical one. In particular, sound waves are, unlike the previous example, of the P-mode.

2.3 Wave Motion in Curvilinear Coordinates and Separability

It has been shown in Section 2.2 that the displacement vector can be recovered from one scalar, Π, and one vector, $\boldsymbol{F}$, potentials obeying the wave equation. The vector wave equation is, in fact, a set of three equations with respect to three components of $\boldsymbol{F}$. Taking into account that according to (2.15) and (2.21)

$$\operatorname{div} \mathbf{curl} \boldsymbol{F} = 0$$

one can reduce the number of independent equations with respect to $\boldsymbol{F}$ to two relations only. The above conclusion holds for any coordinate system. There is, however, the following important feature. In a cartesian reference frame $\nabla^2 \boldsymbol{F}$ splits into three uncoupled equations with respect to F_j,

$$\nabla^2 \boldsymbol{F} \rightarrow \sum_{i=1,2,3} \partial^2 F_j / \partial x_i^2 \ , \quad j = 1,2,3$$

which enables one to develop efficient methods of solution. Unfortunately, it is not the case for curvilinear coordinates, where $\nabla^2 \boldsymbol{F}$ resolves into, generally speaking, coupled relations, rendering the solution difficult or even impossible. Only under special circumstances it is possible to attain the valuable feature of separability for orthogonal curvilinear coordinates. We shall consider this question in more detail in what follows.

For the sake of definitness, we deal with a cylindrical circular frame, r, θ, and z having unit vectors $\boldsymbol{i}_r$, $\boldsymbol{i}_\theta$, and $\boldsymbol{i}_z$ respectively. The potential, $\boldsymbol{F}$, which gives rise to the shear waves, may be represented as a sum of two mutually normal vectors

$$\boldsymbol{F} = \boldsymbol{i}_z \psi + \ell \nabla \times (\boldsymbol{i}_z \chi) \ , \tag{2.26}$$

where $\psi = \psi(r,\theta,z,t)$ and $\chi = \chi(r,\theta,z,t)$ are unknown functions, whereas ℓ is a scalar factor having a dimension of length. The sole role of ℓ is merely to fit units in the right-hand side of the above equation. To satisfy (2.23), the terms appearing in (2.26) must obey the relations

$$c_\beta^2 \nabla^2 (\boldsymbol{i}_z \psi) = \boldsymbol{i}_z \psi_{,tt}$$

$$c_\beta^2 \nabla^2 [\nabla \times (\boldsymbol{i}_z \chi)] = \nabla \times (\boldsymbol{i}_z \chi_{,tt})$$

or

$$c_\beta^2 \nabla^2 \psi = \psi_{,tt} \tag{2.27}$$

$$c_\beta^2 \nabla^2 \chi = \chi_{,tt} \ . \tag{2.28}$$

Recalling also that the dilatational potential Π is governed by (2.22)

$$c_\alpha^2 \nabla^2 \Pi = \Pi_{,tt} \tag{2.29}$$

one arrives at the three uncoupled wave equations with respect to three scalar potentials Π, ψ and χ, namely (2.27) – (2.29).

The associated displacement $\boldsymbol{u}$ may now be represented as follows:

$$\boldsymbol{u} = \boldsymbol{u}_\alpha + \boldsymbol{u}_\beta = \boldsymbol{L} + \boldsymbol{M} + \boldsymbol{N} \ , \tag{2.30}$$

where

$$\boldsymbol{L} = \boldsymbol{u}_\alpha = \nabla \Pi \tag{2.31}$$

$$\boldsymbol{M} = \nabla \times (\boldsymbol{i}_z \psi) \tag{2.32}$$

$$\boldsymbol{N} = \ell \nabla \times \nabla \times (\boldsymbol{i}_z \chi) \tag{2.33}$$

$$M + N = u_\beta \quad . \tag{2.34}$$

The vector M is thus always normal to i_z. More explicit expressions for the displacements will be given in the following sections.

The derivation carried out can be shown valid for two other cylindrical frames too, namely for elliptical and parabolic. Moreover, the representation given by (2.27) – (2.34) may be adapted for the spherical frame as well, provided i_z is replaced by a unit vector in the radial direction and the shear potentials are multiplied by r. In particular cases of symmetrical motion the separability can be achieved for other systems as well. The results obtained provide the basis for a solution to a variety of radiation and diffraction problems and will be of extensive use in the sequel.

2.4 Plane Waves

Because of its simplicity the propagation of plane waves of displacements is of primary interest. Moreover, plane waves serve as convenient building blocks for investigation of other disturbances.

Now we specify some of the above results to the case of plane waves. Denoting the propagation direction by a unit vector n we write a plane wave as

$$u = Af(r \cdot n - Vt) = Af(x_i n_i - Vt) \quad , \tag{2.35}$$

where r is the radius-vector, f an arbitrary function, V the wave velocity, n the wave normal, and A the "amplitude". Insertion of (2.35) in (2.12), provides

$$(\lambda + \mu)(A \cdot n)n + (\mu - \rho V^2)A = 0 \quad .$$

This equation is satisfied in the two cases: i) $A = |A|n$ and $V^2 = c_\alpha^2 = (\lambda + 2\mu)/\rho$ or ii) $A \cdot n = 0$ and $V^2 = c_\beta^2 = \mu/\rho$. Clearly, the first case corresponds to a plane P-wave, while the second one to a plane S-wave. We have thus recovered the result given by (2.8) – (2.10).

Plane waves may also be formulated in terms of potentials. Setting

$$F = 0, \quad \Pi = \Pi_o f(r \cdot n - c_\alpha t)$$

one finds via (2.21)

$$u_\alpha = \operatorname{grad} \Pi = \Pi_o n f'(\tau), \quad u_\beta = 0 \quad , \tag{2.36}$$

where $\tau = r \cdot n - c_\alpha t$ and prime indicates differentiation with respect to the argument.

A shear wave is recovered by introducing a unit vector s, normal to n,

$$s \cdot n = 0 \quad .$$

Setting

$$\Pi = 0, \quad F = \psi_o s \times n f(r \cdot n - c_\beta t)$$

and substituting it in (2.21) we get for S-waves

$$u_\beta = \operatorname{curl} \boldsymbol{F} = \psi_o \boldsymbol{s} f'(\tau), \quad u_\alpha = 0 \tag{2.37}$$

where $\tau = \boldsymbol{r} \cdot \boldsymbol{n} - c_\beta t$.

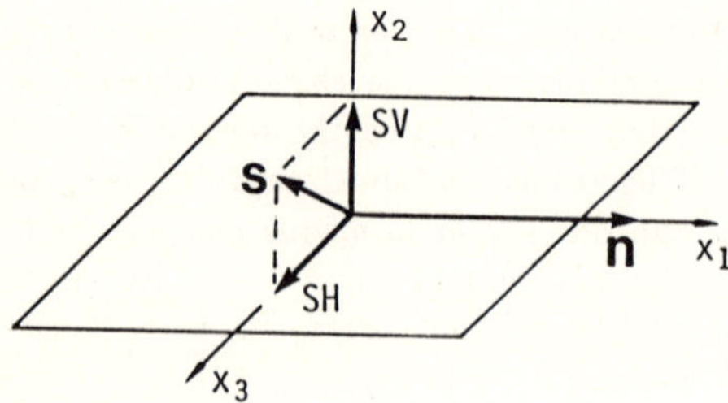

Fig. 2-2 SV- and SH-waves.

Unlike the polarization of P-waves, which is parallel to $\boldsymbol{n}$, uncertainty still exists about the $\boldsymbol{s}$-direction, which may be along any one of the vectors normal to $\boldsymbol{n}$. It is useful to introduce specific notations to identify different shear waves. Figure 2-2 shows that an arbitrary polarization $\boldsymbol{s}$ can be decomposed into vertical (SV) and horizontal (SH) components.

The function f, appearing in the foregoing, may be prescribed according to the nature of the problem at hand. It is convenient to use the wave potentials for this purpose. For example, plane harmonic P-waves are defined by the potentials

$$\begin{aligned} \Pi &= \Pi_o e^{ik_\alpha(\boldsymbol{r}\cdot\boldsymbol{n} - c_\alpha t)} = \Pi_o e^{ik_\alpha \tau} \\ \boldsymbol{F} &= 0, \quad k_\alpha = \omega / c_\alpha \quad , \end{aligned} \tag{2.38}$$

where k_α is the wave number and ω is the circular frequency (radians per second). The associated displacement, $\boldsymbol{u}$, is, according to (2.21),

$$\boldsymbol{u} = \boldsymbol{u}_\alpha = ik_\alpha \Pi_o \boldsymbol{n} e^{ik_\alpha \tau} \quad . \tag{2.39}$$

Although the functions given by (2.38) and (2.39) are complex values (the parameter Π_o is also taken to be complex, say, $\Pi_o = a + ib$), it is understood that either real or imaginary part represents the motion.

For harmonic waves the concept of SV- and SH-components admits the following generalization. Let us introduce mutually normal unit vectors $\boldsymbol{i}_1$ and $\boldsymbol{i}_2$ which are also normal to the x_3-axis. Then, the resultant displacement associated with the shear waves running along the x_3-axis is

$$\boldsymbol{u}_\beta = (a_1 \boldsymbol{i}_1 + a_2 \boldsymbol{i}_2) e^{i(k_\beta x_3 - \omega t)}$$

where $k_\beta = \omega / c_\beta$, a_1 and a_2 are complex amplitudes in the directions $\boldsymbol{i}_1$ and $\boldsymbol{i}_2$ respectively. Varying the ratio a_1/a_2 one alters the trajectories of the particles in the $\boldsymbol{i}_1\boldsymbol{i}_2$-plane. If a_1/a_2 is a real number the particles move along the direct line making the angle $\tan^{-1}(a_2/a_1)$ with the $\boldsymbol{i}_2$-direction. This is the so-called linear polarization. For a_1/a_2 a complex number, the particles move along elliptical trajectories. The circular polarization takes place when $a_1/a_2 = \pm\sqrt{-1}$.

It is usually assumed that time, t, changes from $-\infty$ to $+\infty$ and, thus, the harmonic waves constitute the steady-state motion. For the transient state the motion is initiated at certain times, say, $t = t_o < \infty$. The unit step function, $h(\tau)$, is often used to model a transient phenomenon. Putting

$$f'(\tau) = h(\tau) = \begin{cases} 0 & \tau < 0 \\ 1 & \tau > 0 \end{cases} \tag{2.40}$$

in (2.36) one gets

$$u_\alpha = \Pi_o \boldsymbol{n} h(\boldsymbol{r} \cdot \boldsymbol{n} - c_\alpha t) \tag{2.41}$$

that represents a step plane P-wave. Thus, at a fixed point $\boldsymbol{r} = \boldsymbol{r}_o$ there is no disturbance for the time $t < (\boldsymbol{r}_o \cdot \boldsymbol{n}/c_\alpha)$, while at $t = \boldsymbol{r}_o \cdot \boldsymbol{n}/c_\alpha$ the jump occurs to the magnitude Π_o. Similarly, a delta function

$$\delta(\tau) = h'(\tau) = f'(\tau)$$

can be exploited to describe transient phenomena. In this case the displacement at the site $\boldsymbol{r}_o$ suddenly increases to a very large magnitude at the time $t = \boldsymbol{r}_o \cdot \boldsymbol{n}/c_\alpha$ and rapidly decays thereafter. Of course, more complicated cases of transient behavior may be considered as well.

It should be pointed out that plane waves are mainly a useful idealization. Usually a source produces non-plane waves, which may be viewed as plane only at large distances from it. The presence of inhomogeneities in a real solid distorts further the wave front. The next two sections will deal therefore with the description of non-plane waves, making use of orthogonal curvilinear coordinates.

2.5 Cylindrical Waves

Plane waves, considered in the previous section, can be thought of as emanating from a vibrating infinite plane. Most of the sources radiate a different wave pattern. Consider, for example, a long cylindrical cavity containing a uniformly distributed charge of explosive. Release of energy during explosion leads to propagation of a disturbance with a circular cylindrical front running outward from the cavity. Analysis of problems of such a type dictates the use of cylindrical coordinates shown in Figure 2-3. This section therefore deals with general representations of cylindrical waves.

In circular cylindrical coordinates r, θ, and z, with the unit vectors $\boldsymbol{e}_r$, $\boldsymbol{e}_\theta$ and $\boldsymbol{e}_z$, the relevant operations of vector analysis are given by

$$\begin{aligned} \operatorname{grad} f &= \nabla f = \boldsymbol{e}_r \frac{\partial f}{\partial r} + \boldsymbol{e}_\theta \frac{1}{r}\frac{\partial f}{\partial \theta} + \boldsymbol{e}_z \frac{\partial f}{\partial z} \quad , \\ \operatorname{div} \boldsymbol{F} &= \nabla \cdot \boldsymbol{F} = \frac{\partial (r F_r)}{r \partial r} + \frac{\partial F_\theta}{r \partial \theta} + \frac{\partial F_z}{\partial z} \quad , \end{aligned} \tag{2.42}$$

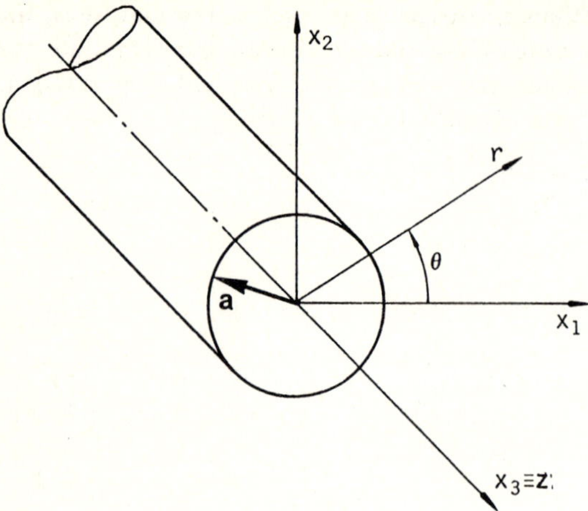

Fig. 2-3 Cylindrical reference frame.

$$\begin{aligned}
\operatorname{curl} \boldsymbol{F} = \nabla \times \boldsymbol{F} &= \boldsymbol{e}_r \frac{1}{r}\left[\frac{\partial F_z}{\partial \theta} - \frac{\partial (rF_\theta)}{\partial z}\right] + \boldsymbol{e}_\theta \left[\frac{\partial F_r}{\partial z} - \frac{\partial F_z}{\partial r}\right] \\
&\quad + \boldsymbol{e}_r \frac{1}{r}\left[\frac{\partial (rF_\theta)}{\partial r} - \frac{\partial F_r}{\partial \theta}\right], \\
\nabla^2 f &= \frac{1}{r}\frac{\partial}{\partial r}\left(r\frac{\partial f}{\partial r}\right) + \frac{1}{r^2}\frac{\partial^2 f}{\partial \theta^2} + \frac{\partial^2 f}{\partial z^2} \quad .
\end{aligned}$$

The scalar equation, governing the potentials Π, ψ, and χ, according to (2.27) - (2.29), is

$$\nabla^2 f = \frac{1}{c^2}\frac{\partial^2 f}{\partial t^2} = \frac{1}{r}\frac{\partial}{\partial r}\left(r\frac{\partial f}{\partial r}\right) + \frac{1}{r^2}\frac{\partial^2 f}{\partial \theta^2} + \frac{\partial^2 f}{\partial z^2} = \frac{1}{c^2}\frac{\partial^2 f}{\partial t^2} \quad , \tag{2.43}$$

where $f = \Pi$, ψ or χ and $c = c_\alpha$ for Π and $c = c_\beta$ for ψ and χ. The solution to this equation for the case of harmonic waves is of primary interest. In fact, having determined that, one can investigate others by making use of the superposition principle.

Setting $f(r,\theta,z,t) = \Phi(r,\theta,z)e^{-i\omega t}$ and inserting this into (2.43) one gets the Helmholtz equation

$$\frac{1}{r}\frac{\partial}{\partial r}\left(r\frac{\partial \Phi}{\partial r}\right) + \frac{1}{r^2}\frac{\partial^2 \Phi}{\partial \theta^2} + \frac{\partial^2 \Phi}{\partial z^2} + \frac{\omega^2}{c^2}\Phi = 0 \quad . \tag{2.44}$$

This can readily be solved by separation of variables. Under the assumption

$$\Phi = R(r)\Theta(\theta)Z(z) \tag{2.45}$$

(2.44) yields

$$\frac{1}{R}\frac{1}{r}\frac{d}{dr}\left(r\frac{dR}{dr}\right) + \frac{1}{\Theta}\frac{1}{r^2}\frac{d^2\Theta}{d\theta^2} + \frac{1}{Z}\frac{d^2 Z}{dz^2} + k^2 = 0 \quad ,$$

where $k = \omega/c$. Since $Z^{-1}d^2Z/dz^2$ depends solely on z and the rest of the terms are independent of z, the above relation provides

$$Z^{-1}d^2Z/dz^2 + k^2 = \mu_1^2 \tag{2.46}$$
$$R^{-1}r^{-1}d(rdR/dr)/dr + r^{-2}\Theta^{-1}d^2\Theta/d\theta^2 = -\mu_1^2 \quad ,$$

where μ_1 is a constant. Multiplying the last expression by r^2 we get two uncoupled equations

$$\Theta^{-1}d^2\Theta/d\theta^2 = -\lambda_1^2 \tag{2.47}$$
$$rR^{-1}d(rdR/dr)/dr + r^2\mu_1 = \lambda_1^2 \quad , \tag{2.48}$$

where λ_1 is a new separation constant. Thus, (2.44) reduces to a set of ordinary differential equations (2.46) – (2.48).

The solution to the Bessel equation (2.48) is given by Bessel functions of the first and second kinds, $J_n(\mu_1 r)$ and $Y_n(\mu_1 r)$, respectively. The alternative is to use the Hankel functions of the first and second kinds, $H_n^{(1)}(\mu_1 r)$ and $H_n^{(2)}(\mu_1 r)$. The first of them describes progressive waves diverging from the z-axis, while the second one those converging on this axis. This becomes clear when one considers the expansions of Hankel functions for large arguments

$$H_n^{(1)(2)}(\mu_1 r)e^{-i\omega t} \simeq \sqrt{2/(\pi\mu_1 r)}e^{\pm i(\mu_1 r-\pi/4-n\pi/2)}e^{-i\omega t} \quad , \tag{2.49}$$

whenever $\mu_1 r \rightarrow \infty$. Note that $H_n^{(1)(2)}(x) = J_n(x) \pm iY_n(x)$.

The solutions to the remaining equations (2.47) and (2.46), are well-known

$$\Theta(\theta) = e^{\pm i\lambda_1\theta} \tag{2.50}$$
$$Z(z) = e^{\pm i(k^2-\mu_1^2)^{1/2}} \quad . \tag{2.51}$$

Usually, $\Theta(\theta)$ must be single valued function, i.e., $\Theta(\theta + 2\pi) = \Theta(\theta)$, which implies λ_1 equals an integer n, $\lambda_1 \equiv n$. The other constant of separation, μ_1, may be complex and is to be specified from additional conditions concerning wave propagation.

Hence, an appropriate typical solution for progressive waves is

$$f_n(r,\theta,z,t) = \left[A_n H_n^{(1)}(\mu_1 r) + B_n H_n^{(2)}(\mu_1 r)\right] e^{\pm in\theta}e^{-i(\omega t \mp \gamma_1 z)} \quad , \tag{2.52}$$

where $\gamma_1^2 = (k^2 - \mu_1^2)$. The coefficients, A_n and B_n, as well as μ_1 and n are to be found from specific conditions of the problem at hand.

For example, if the propagation occurs in the $r\theta$-plane solely, then f_n would be independent of z, which implies

$$\gamma_1 = 0 \quad \rightarrow \quad \mu_1 = k = \omega/c \quad . \tag{2.53}$$

For waves symmetrical about the z-axis no dependence upon θ is present and hence $n = 0$. Thus

$$f_n(r,z,t) = 0, \quad \text{for } n \neq 0$$

$$f_o(r,z,t) = \left[A_o H_o^{(1)}(\mu_1 r) + B_o H_o^{(2)}(\mu_1 r)\right]$$

$$e^{-i(\omega t \mp \gamma_1 z)} \quad . \tag{2.54}$$

Standing (non-propagating) waves are constructed by incorporating the reflected waves along with the incident ones. This leads to the following description of vibrations in cylindrical coordinates

$$f_n(r,\theta,z,t) = [A_n J_n(\mu_1 r) + B_n Y_n(\mu_1 r)] \begin{matrix} \sin n\theta \\ \cos n\theta \end{matrix} \begin{matrix} \sin \gamma_1 z \\ \cos \gamma_1 z \end{matrix} e^{-i\omega t}. \tag{2.55}$$

A relation between plane and cylindrical waves is of substantial interest. To this end, consider a plane wave u travelling along, say, the x_1-direction

$$u = u_o e^{i(kx_1 - \omega t)} = u_o e^{i(kr\cos\theta - \omega t)} \quad , \tag{2.56}$$

where $k = \omega/c$. Since $e^{ikr\cos\theta}$ is periodic in θ, it expands into a Fourier series

$$e^{ikr\cos\theta} = \frac{1}{\sqrt{2\pi}} \sum_{n=-\infty}^{\infty} c_n(r) e^{in\theta} \quad , \tag{2.57}$$

where

$$c_n(r) = \frac{1}{\sqrt{2\pi}} \int_0^{2\pi} e^{ikr\cos\theta} \cos n\theta d\theta \quad .$$

On invoking the integral definition of the Bessel function

$$J_n(kr) = (2\pi i^n)^{-1} \int_0^{2\pi} e^{ikr\cos\theta} \cos n\theta d\theta$$

one gets

$$c_n(r) = i^n J_n(kr)$$

and, hence,

$$e^{ikr\cos\theta} = \sum_{n=-\infty}^{\infty} i^n J_n(kr) e^{in\theta} \quad .$$

Since $J_{-n}(kr) = (-1)^n J_n(kr)$, it finally yields via (2.56) and (2.57) the following remarkable representation of the plane wave as a superposition of the infinite number of cylindrical waves:

$$u = u_o \sum_{n=0}^{\infty} \xi_n i^n J_n(kr) \cos n\theta e^{-i\omega t} \quad , \tag{2.58}$$

where

$$\xi_n = \begin{cases} 1, & \text{n} = 0 \\ 2, & \text{n} = 1,2 \end{cases} \quad .$$

This expression finds extensive use in investigations of diffraction problems involving cylindrical waves.

Having determined the potentials Π, ψ, and χ, one may derive the displacement $\boldsymbol{u}$. In fact, (2.30) – (2.34) provide

$$\begin{aligned} u_r &= \frac{\partial \Pi}{\partial r} + \frac{1}{r}\frac{\partial \psi}{\partial \theta} + \ell \frac{\partial^2 \chi}{\partial r \partial z} \\ u_\theta &= \frac{1}{r}\frac{\partial \Pi}{\partial \theta} - \frac{\partial \psi}{\partial r} + \ell \frac{1}{r}\frac{\partial^2 \chi}{\partial \theta \partial z} \\ u_z &= \frac{\partial \Pi}{\partial z} - \ell \left[\frac{1}{r}\frac{\partial}{\partial r}\left(r\frac{\partial \chi}{\partial r}\right) + \frac{1}{r^2}\frac{\partial^2 \chi}{\partial \theta^2}\right] \quad , \end{aligned} \tag{2.59}$$

which, in turn, yield the stresses with the help of Hooke's law.

2.6 Spherical Waves

Following the same scheme, we consider spherical waves. In spherical coordinates, ϕ, θ, and r, with unit vectors $\boldsymbol{e}_\phi$, $\boldsymbol{e}_\theta$, and $\boldsymbol{e}_r$, shown in Figure 2-4, the relevant differential operations are given by

$$\begin{aligned} \nabla f &= \boldsymbol{e}_r \frac{\partial f}{\partial r} + \boldsymbol{e}_\theta \frac{1}{r}\frac{\partial f}{\partial \theta} + \boldsymbol{e}_\phi \frac{1}{r \sin\theta}\frac{\partial f}{\partial \phi} \\ \nabla \cdot \boldsymbol{F} &= \frac{1}{r^2}\frac{\partial}{\partial r}(r^2 F_r) + \frac{1}{r\sin\theta}\frac{\partial}{\partial\theta}(\sin\theta F_\theta) + \frac{1}{r\sin\theta}\frac{\partial F_\phi}{\partial\phi} \\ \nabla \times \boldsymbol{F} &= \boldsymbol{e}_r \frac{1}{r\sin\theta}\left[\frac{\partial}{\partial\theta}(\sin\theta F_\phi) - \frac{\partial F_\theta}{\partial\phi}\right] + \boldsymbol{e}_\theta \frac{1}{r}\left[\frac{1}{\sin\theta}\frac{\partial F_r}{\partial\phi} - \frac{\partial}{\partial r}(rF_\phi)\right] \\ &\quad + \boldsymbol{e}_\phi \frac{1}{r}\left[\frac{\partial}{\partial r}(rF_\theta) - \frac{\partial F_r}{\partial\theta}\right] \\ \nabla^2 f &= \frac{1}{r^2}\frac{\partial}{\partial r}\left(r^2\frac{\partial f}{\partial r}\right) + \frac{1}{r^2\sin\theta}\frac{\partial}{\partial\theta}\left(\sin\theta\frac{\partial f}{\partial\theta}\right) + \frac{1}{r^2\sin^2\theta}\frac{\partial^2 f}{\partial\phi^2} \quad , \end{aligned}$$

while the scalar wave equation is

$$\nabla^2 f = \frac{1}{c^2}\frac{\partial^2 f}{\partial t^2} \quad . \tag{2.60}$$

The above relation again holds common for $\Pi(c = c_\alpha)$, ψ, and $\chi(c = c_\beta)$.

As has been noted, the solution to (2.60) is of primary interest, since the displacement and stresses follow in a straightforward manner from the wave potentials. Consider the harmonic case, when $f = \Phi(r, \theta, \phi)e^{-i\omega t}$. Inserting this expression into (2.60) one arrives at the Helmholtz equation in spherical coordinates

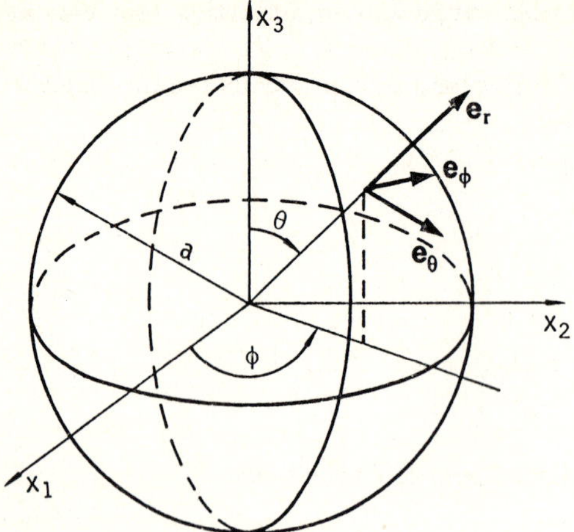

Fig. 2-4 Spherical reference frame.

$$\frac{1}{r^2}\frac{\partial}{\partial r}\left(r^2\frac{\partial\Phi}{\partial r}\right)+\frac{1}{r^2\sin\theta}\frac{\partial}{\partial\theta}\left(\sin\theta\frac{\partial\Phi}{\partial\theta}\right)$$
$$+\frac{1}{r^2\sin^2\theta}\frac{\partial^2\Phi}{\partial\phi^2}+\frac{\omega^2}{c^2}\Phi=0 \quad . \tag{2.61}$$

The method of separation of variables is used once again. On letting $\Phi(r,\theta,\phi)=\varepsilon_1(r)\varepsilon_2(\theta)\varepsilon_3(\phi)$, one finds three ordinary differential equations

$$r^2\frac{d^2\varepsilon_1}{dr^2}+2r\frac{d\varepsilon_1}{dr}+(k^2r^2-d^2)\varepsilon_1=0 \tag{2.62}$$

$$\frac{1}{\sin\theta}\frac{d}{d\theta}(\sin\theta\frac{d\varepsilon_2}{d\theta})+(d^2-\frac{q^2}{\sin^2\theta})\varepsilon_2=0 \tag{2.63}$$

$$\frac{d^2\varepsilon_3}{d\phi^2}+q^2\varepsilon_3=0 \quad . \tag{2.64}$$

Here $k=\omega/c$, while d and q are separation constants. The solution to (2.64) is well known

$$\varepsilon_3(\phi)=e^{\pm iq\phi} \quad .$$

Since only a single-valued function is of interest under regular physical circumstances, $\varepsilon_3(\phi)$ should be periodic in ϕ with period 2π. Thus, q is an integer

$$q=m=0,\ \pm1,\ldots .$$

Making also the substitution $d^2=\nu(\nu+1)$ and $\mu=\cos\theta$ in (2.63), one gets

$$(1-\mu^2)\frac{d^2\varepsilon_2}{d\mu^2}-2\mu\frac{d\varepsilon_2}{d\mu}+\left[\nu(\nu+1)-\frac{m^2}{(1-\mu^2)}\right]\varepsilon_2=0 \quad .$$

This relation can be recognized as the associated Legendre equation. Its two solutions are the Legendre functions, $P_\nu^m(\mu)$ and $Q_\nu^m(\mu)$. The latter is singular at $\mu = \pm 1$ and may be disregarded. To avoid the singularity in the other solution, given by $P_\nu^m(\mu)$, one sets $\nu = n$, where n is an integer.

Turning to (2.62) one makes the substitution $\varepsilon_1(r) = R(r)/\sqrt{kr}$ to arrive at the Bessel equation

$$r^2\frac{d^2R}{dr^2} + r\frac{dR}{dr} + \left[k^2r^2 - (n+\frac{1}{2})^2\right]R = 0.$$

Its solution can be conveniently represented by means of the spherical Bessel functions, $j_n(kr)$, $y_n(kr)$, or $h_n^{(1)(2)}(kr)$, as follows:

$$R(r) = \begin{cases} \sqrt{2kr/\pi} & j_n(kr) \\ \sqrt{2kr/\pi} & y_n(kr) \\ \sqrt{2kr/\pi} & h_n^{(1)(2)}(kr) \end{cases} \quad . \tag{2.65}$$

The choice of an appropriate function is made on the basis of physical considerations of a particular problem at hand.

Thus, a typical spherical wave function is found to be

$$f_n = \varepsilon_1(r)\varepsilon_2(\theta)\varepsilon_3(\phi) = \sqrt{2/\pi}E_n(kr)P_n^m(\mu)e^{\pm im\phi}e^{-i\omega t} \quad ,$$

where $E_n(kr)$ is any one of the above spherical Bessel functions. The $j_n(kr)$ and $y_n(kr)$ are appropriate to standing waves, whereas $h_n^{(1)}(kr)$ or $h_n^{(2)}(kr)$ to travelling disturbances.

In the case of symmetry with respect to x_3 (see Figure 2-4) no dependence on ϕ is present, which implies $m = 0$ and

$$f_n = \sqrt{2/\pi}E_n(kr)P_n(\mu)e^{-i\omega t} \quad , \tag{2.66}$$

with $\mu = \cos\theta$.

It remains to find the representation of plane waves via spherical ones. To this end, consider, say, a P-wave travelling in the x_3-direction.

$$\Pi = \Pi_o e^{i(k_\alpha x_3 - \omega t)} \quad .$$

Since $x_3 = r\cos\theta$, we get

$$\Pi = \Pi_o e^{ik_\alpha r\cos\theta}e^{-i\omega t} \quad . \tag{2.67}$$

Next, we show that

$$e^{ik_\alpha r\cos\theta} = \sum_{n=0}^{\infty} a_n j_n(k_\alpha r)P_n(\cos\theta) \quad , \tag{2.68}$$

with

$$a_n = (2n+1)i^n \quad . \tag{2.69}$$

In fact, multiplying (2.68) by $P_n(\cos\theta)\sin\theta d\theta$ and integrating from 0 to π we get

$$\int_0^{\pi} e^{ik_\alpha r\cos\theta} P_n(\cos\theta)\sin\theta d\theta = \frac{2}{2n+1} a_n j_n(k_\alpha r) \quad ,$$

where the following orthogonality conditions have been invoked

$$\int_{-1}^{1} P_n(\mu)P_\ell(\mu)d\mu = 0, \qquad \text{for } n \neq \ell$$

$$\int_{-1}^{1} P_n^2(\mu)d\mu = \frac{2}{2n+1} \quad ,$$

with $\mu = \cos\theta$. Equations (2.67) – (2.69) express a plane wave in terms of spherical disturbances and are widely used in the analysis of diffraction on spherical surfaces. The displacements u_r, u_θ, and u_ϕ follow from (2.30) – (2.34) and the replacements $\psi \to r\psi$, $\chi \to r\chi$:

$$\begin{aligned} u_r &= \frac{\partial \Pi}{\partial r} + \ell\Big[\frac{\partial^2(r\chi)}{\partial r^2} - r\nabla^2\chi\Big] \\ u_\theta &= \frac{1}{r}\frac{\partial \Pi}{\partial \theta} + \frac{1}{r\sin\theta}\frac{\partial(r\psi)}{\partial \phi} + \ell\frac{1}{r}\frac{\partial^2(r\chi)}{\partial\theta\partial r} \\ u_\phi &= \frac{1}{r\sin\theta}\frac{\partial \Pi}{\partial \phi} - \frac{1}{r}\frac{\partial(r\psi)}{\partial \theta} + \ell\frac{1}{r\sin\theta}\frac{\partial^2(r\chi)}{\partial\phi\partial r} \quad , \end{aligned} \tag{2.70}$$

which enable us to find the strains and stresses in a straightforward way. Having obtained general formal relations describing typical wave patterns, we go over to their applications to physically meaningful problems.

2.7 Radiation from a Cylindrical Cavity

Considering a cylindrical cavity located in an infinitely extended medium (Figure 2-3), we assume that a uniform harmonic pressure acts on the cavity surface, $r = a$, according to the law

$$\sigma_{rr}(r = a, t) = -p_o e^{-i\omega t} \quad , \tag{2.71}$$

with p_o being a known constant. The apparent symmetry of the problem indicates that only the radial displacement, u_r, is not zero and is a spatial function of r solely. Accordingly, no shear waves are present ($\psi = \chi = 0$) and (2.59) yield

$$u_r = \partial\Pi/\partial r \quad , \tag{2.72}$$

where the dilatational wave potential, $\Pi(r,t)$, is governed by the wave equation

$$c_\alpha^2 \nabla^2 \Pi = \Pi,_{tt} \quad . \tag{2.73}$$

The appropriate solution to (2.73) is given by (2.54), where γ_1 must be set to zero, since no dependence on z is present. Accordingly, $\mu_1 = k_\alpha = \omega/c_\alpha$, and

$$\Pi = \left[A_o H_o^{(1)}(k_\alpha r) + B_o H_o^{(2)}(k_\alpha r)\right] e^{-i\omega t} \quad . \tag{2.74}$$

To specify A_o and B_o one applies the following physical consideration. First, since the internal pressure, applied at $r = a$, excites outgoing waves solely, the last term in (2.74) must be suppressed, which implies $B_o = 0$. Second, the boundary condition, given by (2.71), enables one to identify the remaining coefficient A_o. Hooke's law provides

$$\sigma_{rr} = \lambda(\varepsilon_{rr} + \varepsilon_{\theta\theta} + \varepsilon_{zz}) + 2\mu\varepsilon_{rr} \quad .$$

Since $\varepsilon_{zz} = 0$, $\varepsilon_{\theta\theta} = u_r/r$, and $\varepsilon_{rr} = \partial u_r/\partial r$, it yields

$$\sigma_{rr} = \lambda(\partial u_r/\partial r + u_r/r) + 2\mu\partial u_r/\partial r \quad .$$

Making use of (2.72) - (2.74) the stress is found to be

$$\sigma_{rr} = \left[-(\lambda + 2\mu)k_\alpha^2 H_o^{(1)}(k_\alpha r) - (2\mu/r)\frac{d}{dr}H_o^{(1)}(k_\alpha r)\right] A_o e^{-i\omega t} \quad . \tag{2.75}$$

Taking into account that

$$\frac{d}{dr}H_o^{(1)}(k_\alpha r) = -k_\alpha H_1^{(1)}(k_\alpha r) \quad ,$$

substituting a for r in (2.75), and equating the result to $-p_o e^{-i\omega t}$, one finally finds

$$A_o = p_o\left[(\lambda + 2\mu)k_\alpha^2 H_o^{(1)}(k_\alpha a) - 2\mu k_\alpha H_1^{(1)}(k_\alpha a)/a\right]^{-1} \quad . \tag{2.76}$$

This solution represents a steady-state response of the medium. To investigate its transient behavior the Fourier superposition method can be adopted. The case of an instant explosion of a uniformly distributed cylindrical charge is of particular interest. An appropriate pressure function, applied at $r = a$, is given by

$$\sigma_{rr} = -p_o\delta(t) \quad .$$

By making use of the theory of linear systems, the solution to this problem, u_r^δ, could be stated as the Fourier transform of the harmonic response just obtained, in accord with (1.106)

$$u_r^\delta = -\frac{1}{2\pi}\int_{-\infty}^{\infty} A_o(\omega) H_1^{(1)}(\omega r/c_\alpha)(\omega/c_\alpha)e^{-i\omega t}d\omega \quad ,$$

where A_o is defined by (2.76), with $k_\alpha = \omega/c_\alpha$. A similar procedure holds for any other time-dependent pressure. However, evaluation of the integral involved may be rather complicated.

The admittance function of the system, which may be defined as $H(r, i\omega) = H^r(r,\omega) + iH^i(r,\omega) = \sigma_{rr}(r)/\sigma_{rr}(r = a)$, is seen to be complex. This is a manifestation of the specific type of energy loss known as radiation damping, which will be considered in more detail in the next section.

2.8 Radiation from a Rigid Embedded Cylinder. Radiation Damping

The example we consider is a rigid cylinder of infinite length embedded in an elastic medium (Figure 2-3). When the cylinder executes vibrations along the z-axis it is accompanied by radiation of outgoing shear waves, unlike the previous case. Assume that the cylinder motion, u_c, is given by

$$u_c = u_c^o e^{-i\omega t} \tag{2.77}$$

and that there is no relative motion at the boundary surface, $r = a$ (welded contact). Thus

$$u_z(r = a, \theta, z, t) = u_c^o e^{-i\omega t} \quad , \tag{2.78}$$

where u_z is the displacement of the surrounding medium.

The solution to this problem can be found in the class of SH-waves, considered in Section 2.2. We set

$$u_r = u_\theta = 0 \qquad \text{and} \qquad u_z = u_z(r, t) \quad .$$

Observing (2.59), one finds that the above relations for the displacements could be satisfied, if the wave potentials are

$$\Pi = \psi = 0 \qquad \text{and} \qquad \chi = \chi(r, t) \quad .$$

The solution to χ, given by (2.54), is again appropriate, if $\gamma_1 = 0$, $\mu_1 = k_\beta = \omega/c_\beta$ and $B_o = 0$. Thus

$$\chi = A_o H_o^{(1)}(k_\beta r) e^{-i\omega t} \quad ,$$

which differs from the potential of the previous problem only by the wave number. The last of (2.59) yields

$$\begin{aligned} u_z &= -\frac{1}{r}\frac{\partial}{\partial r}\left(r\frac{\partial \chi}{\partial r}\right) \\ &= -A_o\left[k_\beta r \frac{d}{d(k_\beta r)} H_o^{(1)}(k_\beta r) + k_\beta^2 r^2 \frac{d^2}{d(k_\beta r)^2} H_o^{(1)}(k_\beta r)\right] e^{-i\omega t}/r^2 \quad . \end{aligned}$$

Making use of the recurrence formulas

$$dH_o(q)/dq = H_{-1}(q)$$

and

$$d^2 H_o(q)/dq^2 = -H_o(q) - H_{-1}(q)/q$$

the displacement reduces to

$$u_z = k_\beta^2 A_o H_o^{(1)}(k_\beta r) e^{-i\omega t} \quad .$$

The condition, given by (2.78), provides

$$A_o = u_c^o / H_o^{(1)}(k_\beta a) k_\beta^{-2} \quad .$$

Hence, the radiated SH-waves are given by

$$u_z = u_c^o H_o^{(1)}(k_\beta r)/H_o^{(1)}(k_\beta a)e^{-i\omega t}. \tag{2.79}$$

The above procedure intends to illustrate a process of obtaining a solution by making use of potentials. In fact, (2.79) could be deduced immediately from (2.25).

A surprising phenomenon is observed in both of the problems considered: despite the fact that the medium is perfectly elastic, the whole system possesses a specific type of damping. In fact, let us compute the admittance function, $H(r, i\omega)$, for, say, the case of an embedded rigid cylinder. Considering u_c as an excitation and u_z as a response one has

$$H(r, i\omega) = u_z/u_c = H_o^{(1)}(k_\beta r)/H_o^{(1)}(k_\beta a) \quad , \tag{2.80}$$

with

$$\begin{aligned}\mathrm{Re}\,[H(r, i\omega)] = {} & [J_o(k_\beta r)J_o(k_\beta a) + Y_o(k_\beta r)Y_o(k_\beta a)] \\ & /[J_o^2(k_\beta a) + Y_o^2(k_\beta a)]\end{aligned}$$

and

$$\begin{aligned}\mathrm{Im}\,[H(r, i\omega)] = {} & [-J_o(k_\beta r)Y_o(k_\beta a) + Y_o(k_\beta r)J_o(k_\beta a)] \\ & /[J_o^2(k_\beta a) + Y_o^2(k_\beta a)] \quad .\end{aligned}$$

It is known from the theory of linear systems that the imaginary part of the admittance function is intimately related to the energy loss. The phenomenon, which takes place, can easily be appreciated by taking into account that, due to the unboundedness of the medium, the radiated waves would never be reflected. It is called radiation damping.

One of the manifestations of radiation damping is a phase shift between the excitation and the response. Compute, say, the imaginary parts of u_c and u_z which represent the actual motions of the cylinder and the medium, respectively. By means of (2.77), (2.79) and (2.80) we obtain

$$\mathrm{Im}\,[u_c] = -u_c^o \sin\omega t \tag{2.81}$$

$$\begin{aligned}\mathrm{Im}\,[u_z] &= u_c^o\{\mathrm{Im}\,[H(r, i\omega)]\cos\omega t - \mathrm{Re}\,[H(r, i\omega)]\sin\omega t\} \\ &= u_c^o\tilde{A}\sin(\omega t + \tilde{\alpha}) \quad ,\end{aligned} \tag{2.82}$$

where

$$\tilde{A}^2 = |H(r, i\omega)|^2$$

$$\tan\tilde{\alpha} = -\mathrm{Im}\,[H(r, i\omega)]/\,\mathrm{Re}\,[H(r, i\omega)] \quad .$$

Radiation damping constitutes an inherent feature of any phenomenon involving emission of travelling disturbances in an infinite medium.

2.9 Radiation from a Spherical Cavity. Remarks on Radiation Conditions

As our next typical example, consider a spherical cavity of radius a, located in an unbounded elastic medium, and subjected to the pressure $p_o e^{-i\omega t}$. Then

$$\sigma_{rr}(r = a) = -p_o e^{-i\omega t} \quad , \tag{2.83}$$

which causes the disturbances to run outward. Due to the apparent symmetry one may set

$$u_r = u_r(r, t), \qquad u_\phi = u_\theta = 0 \quad .$$

Clearly, this is a case of outgoing P-waves, since the directions of propagation and displacement coincide. Hence, the wave potentials are

$$\Pi = \Pi(r, t), \qquad \chi = \psi = 0 \quad .$$

The solution to Π can be given by a superposition of functions f_n, defined by (2.66). To avoid dependence on $\mu = \cos\theta$, one must put $n = 0$ ($P_o(\mu) = 1$). Also, setting in (2.66)

$$E_o(kr) = h_o^{(1)}(k_\alpha r)$$

to describe outgoing waves, one has

$$\Pi = A_o h_o^{(1)}(k_\alpha r) e^{-i\omega t} \quad . \tag{2.84}$$

Equations (2.70) and Hooke's law, adapted to the case of spherical symmetry, provide

$$u_r = \frac{\partial \Pi}{\partial r}$$

$$\sigma_{rr} = \lambda[\partial^2\Pi/\partial r^2 + 2(\partial\Pi/\partial r)/r] + 2\mu\partial^2\Pi/\partial r^2 \quad .$$

Since Π obeys the reduced wave equation,

$$\partial^2\Pi/\partial r^2 + 2(\partial\Pi/\partial r)/r = \Pi,_{tt}/c_\alpha^2 \quad ,$$

the stress, σ_{rr}, is given by

$$\begin{aligned}\sigma_{rr} &= -(\lambda + 2\mu)k_\alpha^2\Pi - 4\mu(\partial\Pi/\partial r)/r \\ &= -\left[(\lambda + 2\mu)k_\alpha^2 r^2 h_o^{(1)}(k_\alpha r) - 4\mu k_\alpha r h_1^{(1)}(k_\alpha r)\right] A_o e^{-i\omega t}/r^2 \quad ,\end{aligned} \tag{2.85}$$

where the relation

$$dh_o^{(1)}(q)/dq = -h_1^{(1)}(q)$$

has been utilized.

Making use of the boundary condition given by (2.83) and substituting a for r into (2.85) one finds that A_o is as follows:

$$A_o = p_o a^2/Q(k_\alpha a) \quad , \tag{2.86}$$

where

$$Q(k_\alpha a) = \left[(\lambda + 2\mu)k_\alpha^2 a^2 h_o^{(1)}(k_\alpha a) - 4\mu k_\alpha a h_1^{(1)}(k_\alpha a)\right] \quad . \tag{2.87}$$

The final expressions can be simplified by noting that

$$h_o^{(1)}(k_\alpha r) = -ie^{ik_\alpha r}/(k_\alpha r)$$

$$h_1(k_\alpha r) = ie^{ik_\alpha r}(ik_\alpha r - 1)/(k_\alpha r)^2 \quad .$$

Making use of these relations σ_{rr} is found, according to (2.85) - (2.87), to get the following symmetric form:

$$\sigma_{rr} = -\frac{p_o a^3\left[(\lambda + 2\mu)k_\alpha^2 r^2 + 4\mu(ik_\alpha r - 1)\right]}{r^3\left[(\lambda + 2\mu)k_\alpha^2 a^2 + 4\mu(ik_\alpha a - 1)\right]} e^{ik_\alpha(r-a)} e^{-i\omega t} \quad . \tag{2.88}$$

In the radiation problems considered, the Hankel functions of the first kind have been systematically chosen to represent the outgoing diverging waves. It should be pointed out once again that the Hankel functions of the second kind constitute a solution to the wave equation as well, and were omitted on the basis of physical considerations.

An analytical criterion which would allow one a correct choice of functions seems desirable. To this end, consider, say, dilatational potentials of two spherically-symmetric waves, Π_1 and Π_2

$$\Pi_1 \simeq (e^{ik_\alpha r}/r)e^{-i\omega t}, \qquad \Pi_2 \simeq (e^{-ik_\alpha r}/r)e^{-i\omega t} \quad . \tag{2.89}$$

Simple calculations show that

$$\begin{aligned} r(\partial\Pi_1/\partial r - ik_\alpha\Pi_1) &= -\Pi_1 \\ r(\partial\Pi_2/\partial r + ik_\alpha\Pi_2) &= -\Pi_2 \quad . \end{aligned} \tag{2.90}$$

Hence, the potentials behave in different ways for $r \to \infty$. In fact, since $\lim_{\omega\to\infty} \Pi_i = 0$, $(i = 1, 2)$, we get for Π_1

$$\lim_{r\to\infty} [r(\partial\Pi_1/\partial r - ik_\alpha\Pi_1)] = 0 \quad , \tag{2.91}$$

while for Π_2 we similarly obtain

$$\lim_{r\to\infty} [r(\partial\Pi_2/\partial r + ik_\alpha\Pi_2)] = 0 \quad . \tag{2.92}$$

The above relations, known as the radiation conditions, provide an analytical criterion for distinguishing between disturbances which converge or diverge at infinity. They can be extended to an arbitrary type of three-dimensional elastic waves.

2.10 Radiation from a Rigid Embedded Sphere. Imperfect Bonding

We return to interactions of an embedded rigid inclusion with a surrounding elastic medium. This coupled system constitutes a natural generalization of the mass-spring model, well-known in the theory of vibrations, and deriving a solution is of basic interest for elastodynamics as well as for understanding the behavior of composites.

2.10.1 Torsional Oscillations

Let μ and λ be the Lame constants of the matrix and ρ its density. The inclusion density is ρ_o and its radius is a (Figure 2-4). We first consider torsional oscillations of the sphere about the x_3-axis, given by a rotation angle $\phi(t)$, which excite outgoing disturbances. Due to the symmetry involved, it is natural to assume that these disturbances are of pure shear motion associated with the displacement u_ϕ, which depends on r, θ, and t only.

Thus, we assume that the displacement vector, u, is given by

$$u = (0, 0, u_\phi) \qquad \text{and} \qquad u_\phi = u_\phi(r, \theta, t)$$

It follows from (2.70), which present the displacement in terms of the wave potentials, that we may put

$$\Pi = \chi = 0, \qquad \psi = \psi(r, \theta, t) \quad . \tag{2.93}$$

Assuming welded contact, at the surface $r = a$ the matrix displacement is equal to that of the inclusion. Hence, as Figure 2-5 shows,

$$u_\phi(r = a, \theta, t) = \phi(t) a \sin\theta \quad . \tag{2.94}$$

Setting $\phi(t) = de^{-i\omega t}$, one can represent the general solution to ψ as a superposition of the expressions following from (2.66) and the radiation conditions

$$\psi = \sqrt{2/\pi} \sum_{n=0}^{\infty} A_n h_n^{(1)}(k_\beta r) P_n(\cos\theta) e^{-i\omega t} \quad ,$$

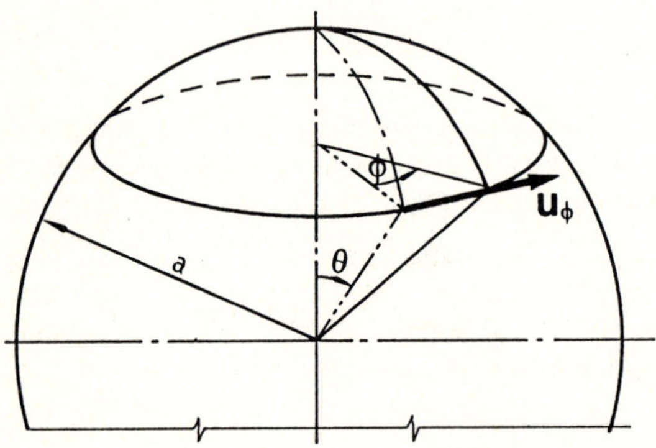

Fig. 2-5 Torsional oscillations of an embedded sphere.

where A_n are as yet unknown coefficients. As (2.70) shows

$$u_\phi = -\frac{1}{r}\partial(r\psi)/\partial\theta = -\partial\psi/\partial\theta \quad .$$

Hence,

$$u_\phi = \sqrt{2/\pi}\sum_{n=0}^{\infty} A_n h_n^{(1)}(k_\beta r)(dP_n(\mu)/d\mu)\sin\theta e^{-i\omega t} \quad . \tag{2.95}$$

Comparing (2.94) and (2.95) and taking into account that

$$P_o(\mu) = 1, \qquad P_1(\mu) = \mu, \qquad P_2(\mu) = (3\mu^2 - 1)/2 \ ...$$

$$(\mu = \cos\theta)$$

one arrives at the conclusion that the suitable values of A_n are

$$A_n = 0 \qquad \text{for} \qquad n \neq 1$$

$$A_1 = da/\left[h_1^{(1)}(k_\beta a)\sqrt{2/\pi}\right] \quad .$$

Since

$$h_1^{(1)}(k_\beta r) = i(ik_\beta r - 1)e^{ik_\beta r}/(k_\beta r)^2 \quad ,$$

the final expression for u_ϕ simplifies to

$$u_\phi = \frac{da^3(ik_\beta r - 1)}{r^2(ik_\beta a - 1)}\sin\theta e^{ik_\beta(r-a)}e^{-i\omega t}. \tag{2.96}$$

Now we can deduce the torque, $G(t)$, which must be applied to the sphere in order to maintain its harmonic oscillations. From the angular momentum theorem one gets

$$G(t) = -\int_0^\pi 2\pi a(a\sin\theta)^2\sigma_{r\phi}(r = a)d\theta + Id^2\phi/dt^2 \quad , \tag{2.97}$$

with I the sphere moment of inertia. The integral term is, of course, the torque due to the only non-zero elastic stress, $\sigma_{r\phi}$. Hooke's law in spherical coordinates yields

$$\sigma_{r\phi} = \mu(\partial u_\phi/\partial r - u_\phi/r) \quad .$$

Inserting (2.96) into this relation and evaluating the integral appearing in (2.97), one gets

$$G(t) = 8\pi da^3\mu G_o\cos(\omega t + \alpha_o)/3 \quad , \tag{2.98}$$

with the abbreviations

$$G_o = \left\{\eta^6 + \left[3 + (2 - k_o)\eta^2 - k_o\eta^4\right]^2\right\}^{1/2}/(1 + \eta^2) \tag{2.99}$$

$$\alpha_o = \tan^{-1}\left\{\eta^3/\left[3 + (2 - k_o)\eta^2 - k_o\eta^4\right]\right\} \tag{2.100}$$

$$k_o = 3I/(8\pi\rho a^5), \qquad \eta = a\omega/c_\beta = a\omega\rho^2/\mu^2 \quad .$$

Taking into account that the real part of the excitation, $\phi(t)$, is given by

$$\mathrm{Re}\,[\phi(t)] = d\cos\omega t$$

and comparing it with (2.98) one observes, as in the previous example, the presence of the phase lead, α_o, due to radiation damping. The functions $G_o = G_o(\eta)$ and $\alpha_o = \alpha_o(\eta)$, revealing the effect of frequency on the amplitude and phase of the radiated field, are shown in Figures 2-6 and 2-7, according to (2.99) and (2.100).

2.10.2 Rectilinear Oscillations

Now we turn to the case of rectilinear oscillations, assuming that the rigid sphere executes a motion along the x_3-axis given by

$$u_3 = u_o e^{-i\omega t} \quad ,$$

which gives rise to dilatational and shear waves in a surrounding medium (see Figure 2-4). The procedure follows that of the previous problem, but is far more involved. Because of the symmetry, $u_\phi = 0$, and the displacement vector, u, is

$$u = (u_r, u_\theta, 0)$$

where $u_r = u_r(r,\theta,t)$ and $u_\theta = u_\theta(r,\theta,t)$.

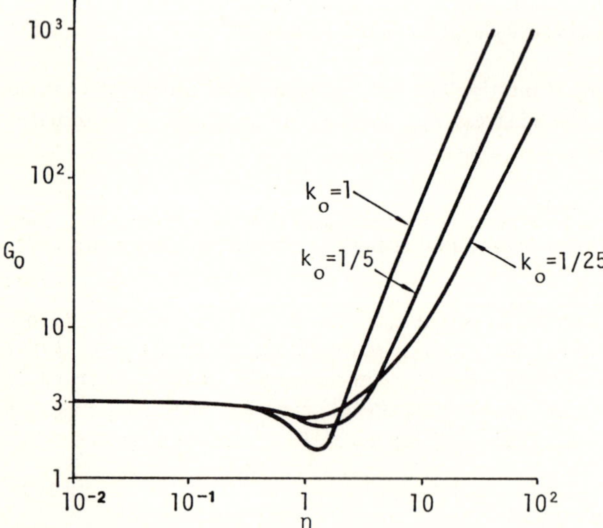

Fig. 2-6 Amplitude vs. frequency.

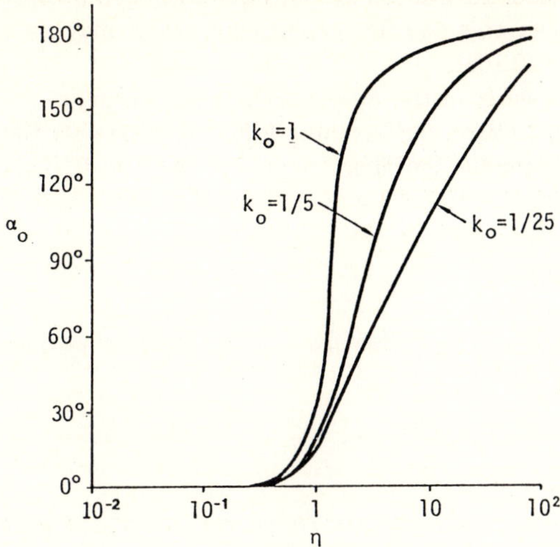

Fig. 2-7 Phase lead vs. frequency.

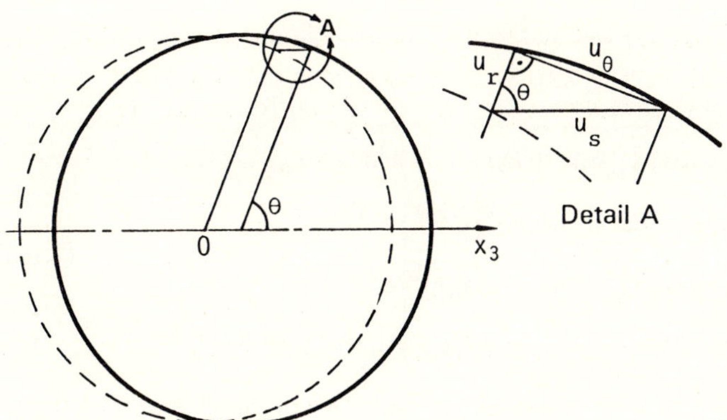

Fig. 2-8 Rectilinear oscillations of an embedded sphere.

Referring to Figure 2-8, the boundary conditions for the welded contact are

$$\begin{aligned} u_r(r = a, \theta, t) &= u_s \cos\theta \\ u_\theta(r = a, \theta, t) &= -u_s \sin\theta \quad . \end{aligned} \tag{2.101}$$

It should be pointed out that an actual interfacial contact is not always properly reflected by (2.101). For example, in composite materials, natural or man-made, imperfect bonding may occur, which allows for a certain amount of relative motion between the phases. In this case, the boundary conditions may be given by

$$\begin{aligned} u_r(r = a, \theta, t) &= u_s \cos\theta \\ \sigma_{r\theta}(r = a, \theta, t) &= 0 \quad , \end{aligned} \tag{2.102}$$

which allow for a slip at the interface. To account for both cases we shall present a general form of the solution containing two free coefficients, which may then be specified either from (2.101) or (2.102).

Unlike the previous case, the shear in the $\phi\theta$-plane does not occur, so we may put the proper potential, ψ, to zero. On invoking (2.66) to represent the remaining potentials, Π and χ, we try the functions

$$\Pi = A_1 h_1^{(1)}(k_\alpha r)\cos\theta e^{-i\omega t}$$

$$\chi = B_1 h_1^{(1)}(k_\beta r)\cos\theta e^{-i\omega t} \quad .$$

The associated displacement, u_r and u_θ, follow from (2.70) and may be conveniently expressed in elementary functions as

$$\begin{aligned} u_r &= N_1\cos\theta\Big\{\Big[2-2ik_\alpha r-(k_\alpha r)^2\Big]e^{ik_\alpha r} \\ &\quad + 2N_2(ik_\beta r-1)e^{ik_\beta r}\Big\}e^{-i\omega t}/r^3 \\ u_\theta &= -N_1\sin\theta\Big\{(ik_\alpha r-1)e^{ik_\alpha r} \\ &\quad + N_2\Big[1-ik_\beta r-(k_\beta r)^2\Big]e^{ik_\beta r}\Big\}e^{-i\omega t}/r^3. \end{aligned} \tag{2.103}$$

The coefficients N_1 and N_2 can be specified from the boundary conditions given by (2.101) or (2.102).

Accordingly, the stresses are

$$\begin{aligned} \sigma_{rr} &= -N_1\mu\cos\theta\Big\{[(k_\beta r)^2(ik_\alpha r-1)+4\Big[3-3ik_\alpha r-(k_\alpha r)^2\Big]e^{ik_\alpha r} \\ &\quad -4N_2\Big[3-3ik_\beta r-(k_\beta r)^2\Big]e^{ik_\beta r}\Big\}e^{-i\omega t}/r^4 \\ \sigma_{r\theta} &= -N_1\sin\theta\Big\{2\Big[3-3ik_\alpha r-(k_\alpha r)^2\Big]e^{ik_\alpha r}- \\ &\quad N_2\Big[6-6ik_\beta r-3(k_\beta r)^2+i(k_\beta r)^3\Big]e^{ik_\beta r}\Big\}e^{-i\omega t}/r^4 \end{aligned} \tag{2.104}$$

which completes the solution, provided N_1 and N_2 are found from (2.101) or (2.102). The force, F, which must be applied to the sphere to maintain its steady vibrations, may be obtained from the equation of motion

$$\begin{aligned} &-F+(4\pi a^3/3)\rho_0\ddot{u}_s \\ &= \int_0^\pi 2\pi a(a\sin\theta)(\sigma_{rr}\cos\theta-\sigma_{r\theta}\sin\theta)d\theta\,\Big|_{r=a} \quad , \end{aligned} \tag{2.105}$$

which is left to exercises. Its explicit expression for the important case of low frequencies will be given in the next section.

2.11 Multipoles, Resonances, and Related Considerations

The above solutions for radiation from cavities and rigid inclusions hold for any ratio of the inclusion size to the wavelength, a/λ. In some cases the inclusion is small comparable to the wavelength, which motivates a search for more simple results by investigating the case of the long wavelength, $a/\lambda \ll 1$. Obviously, these are natural extensions of static investigations, which may, therefore, be capable of displaying dynamic effects in an easily interpretable way.

Another case, which admits simplifications, is the so-called far-field radiation given by the limit $r \to \infty$. The far-field expansion provides a physically meaningful information, when one is not interested in (or is not capable of) treating proximities of a source.

Consider an example of small spherical cavity, which radiates harmonic waves. On assuming $k_\alpha a < 1$, $r/a > 1$, we get from (2.88)

$$\sigma_{rr} = -p_o a^3 k_\alpha^2 c_\alpha^2 / (4 r c_\beta^2) e^{i(k_\alpha r - \omega t)} \quad . \tag{2.106}$$

The product $(4\pi a^3/3)p_o = D$ may be referred to as the source strength. Then (2.106) yields a simple source of dilatational waves in elastic media, or a monopole

$$\sigma_{rr} = -D 3 k_\alpha^2 c_\alpha^2 / (16 r \pi c_\beta^2) e^{i(k_\alpha r - \omega t)} \quad , \tag{2.107}$$

which resembles a similar concept in the acoustics of fluids.

Note that the stress has infinite value at $r = 0$. However, r is bounded by the above constraint, $r/a > 1$, so the stress never becomes singular, even in the framework of pure elasticity. If the source point is located at $\boldsymbol{r}_o$, then r in (2.107) must be replaced by $|\boldsymbol{r} - \boldsymbol{r}_o|$. It is seen that the stress at $\boldsymbol{r}_o$ caused by a source at $\boldsymbol{r}$ equals the stress at $\boldsymbol{r}$ caused by a source at $\boldsymbol{r}_o$, which is known as the principle of reciprocity.

Now we shall consider rectilinear oscillations of an embedded rigid sphere, which will lead us to the concept of a dipole. For $k_\beta a < 1$ (which implies $k_\alpha a < 1$) we get from (2.101), and (2.103) the values of the coefficients, N_1 and N_2, for the case of perfect bonding

$$N_1 = -\frac{3a^3 u_o}{2\beta^2 + \alpha^2}\left(1 - i\frac{2\beta^3 + \alpha^3}{2\beta^2 + \alpha^2}\right)$$

$$N_2 = \frac{3 - 3i\alpha - \alpha^2}{3 - 3i\beta - \beta^2} e^{i(\alpha - \beta)} \quad ,$$

with $\alpha = k_\alpha a$, $\beta = k_\beta a$, and u_o the amplitude of the sphere oscillations.

Then (2.105) yields the force, F, needed to maintain vibrations

$$F/u_o = \frac{12\pi a^3 \rho \omega^2}{2\beta^2 + \alpha^2}\left(1 - i\frac{2\beta^3 + \alpha^3}{2\beta^2 + \alpha^2}\right) - \frac{4\pi a^3}{3} \rho_o \omega^2 \quad . \tag{2.108}$$

This expression may also be viewed as the dynamic complex stiffness of the system. In particular, the static stiffness, χ_{st}, is

$$\chi_{st} = \lim_{\omega \to 0} (F/u_o) = 12\pi a \mu / (c_\beta^2 / c_\alpha^2 + 2). \tag{2.109}$$

A small oscillating sphere can be, somewhat loosely, called a dipole, similarly to the acoustics of fluids. However, in the case of solid media both, dilatational and shear waves are present. Also, the boundary conditions can be of various types. In fact, a similar procedure applies to an "imperfectly bonded" dipole.

One may also derive a hierarchy of the sources (multipoles) by superimposing radiation due to monopoles or dipoles.

Finally, we turn to discussions on the resonance phenomena and free vibrations. Consider, for example, the case of the steady-state radiation from a spherical cavity. According to (2.88) the stress, σ_{rr}, can be written as

$$\sigma_{rr} = -p_o H(r, i\omega)e^{-i\omega t} \quad , \tag{2.110}$$

where the admittance function, $H(r, i\omega) = H^r(r,\omega) + iH^i(r,\omega)$, is

$$H(r, i\omega) = \frac{a^3\Big[(\lambda + 2\mu)k_\alpha^2 r^2 + 4\mu(ik_\alpha r - 1)\Big]}{r^3\Big[(\lambda + 2\mu)k_\alpha^2 a^2 + 4\mu(ik_\alpha a - 1)\Big]} e^{ik_\alpha(r-a)} \quad . \tag{2.111}$$

The resonance radiation is obviously associated with the roots of the denominator, which are given by

$$(k_\beta a)^2 - 4 + i4(k_\alpha a) = 0 \quad . \tag{2.112}$$

The last term of this expression is due to radiation damping and gives rise to complex roots. In the case of a rubberlike material, $\nu \simeq 0.5$, this term may be neglected, since $c_\alpha \gg c_\beta$. This provides

$$(k_\beta a)_{res} = (\omega a/c_\beta)_{res} \simeq 2$$

and shows a possibility of a strong resonance (low radiation damping), if $\nu \simeq 0.5$. The roots for an arbitrary Poisson's ratio are

$$(\omega a)^{1,2}_{res} = c_\alpha(1 - 2\nu)^{1/2}/(1 - \nu)\Big[\pm 1 - i(1 - 2\nu)^{1/2}\Big]. \tag{2.113}$$

In the case of free vibrations a source radiates a disturbance decaying in time because of radiation damping, which implies the complex frequency of vibrations, $\omega = \omega_r + i\omega_i$. The treatment of free vibrations may be facilitated by an analysis of the response to the unit impulse. In the above case of cavity this is given by

$$\sigma_{rr}(r,t) = -\int_{-\infty}^{\infty} H(r, i\omega)e^{-i\omega t} d\omega/(2\pi) \tag{2.114}$$

with $H(r, i\omega)$ defined by (2.111). Details of this analysis will be given later. The problem simplifies, if one is interested in the natural frequencies and the decay time only. These may be found directly from (2.110) and (2.111) by considering the limiting case $p_o \to 0$. In fact, a non-trivial solution, $\sigma_{rr} \neq 0$, to (2.110) may still exist provided that the denominator in (2.111) vanishes. This yields again (2.113), as expected. Accordingly, "frequency" of radiation is given by ω_r, while time decay is governed by ω_i,

$$e^{-i\omega t} = e^{-i(\omega_r + i\omega_i t)} = e^{\omega_i t} e^{-i\omega_r t}$$

More details will be given in Section 2.15.

2.12 Green's Dynamic Tensor

The approach adopted in the preceding sections makes use of a homogeneous wave equation

$$c\nabla^2 f - \partial^2 f/\partial t^2 = 0 \quad ,$$

which describes free waves, while a physical cause of the phenomenon lies in the boundary and initial conditions. Unlike this, the relation

$$c\nabla^2 f - \partial^2 f/\partial t^2 = -f_o \quad , \tag{2.115}$$

known as a nonhomogeneous wave equation, involves directly the forcing term $f_o \neq 0$, which depends on time and space. It is possible to show that the equations of motion (1.15) reduce to a set of relations of the above type, provided body forces are preserved.

This section deals with waves caused by variable body forces applied to a point of the medium. This problem is of fundamental interest for the linear theory. The method of integral transforms is adopted here, while other techniques are known to provide the result as well.

We first write down (2.3) as

$$L_{i\ell} u_\ell = 0 \quad , \tag{2.116}$$

where the operator, $L_{i\ell}$, is

$$L_{i\ell} = -\delta_{i\ell}\rho\frac{\partial^2}{\partial t^2} + (\lambda+\mu)_{,i\ell} + \mu\nabla^2\delta_{i\ell} \quad . \tag{2.117}$$

Upon the presence of a forcing term, f_i, one gets, instead of (2.116),

$$L_{i\ell} u_\ell = -f_i \quad . \tag{2.118}$$

A specific case of forcing we are interested in is the unit impulse located at the origin of a reference frame and directed along a coordinate x_j, which can be written as $\delta_{ij}\delta(\boldsymbol{r})\delta(t)$. The displacement along x_ℓ occurring at a point $\boldsymbol{r} = (x_1, x_2, x_3)$ due to the unit impulse directed along the x_j-axis will be labelled as $G_{\ell j}(\boldsymbol{r}, t)$. Making these substitutions for the forcing term and displacements in (2.118) one gets

$$L_{i\ell} G_{\ell j}(\boldsymbol{r}, t) = -\delta_{ij}\delta(\boldsymbol{r})\delta(t) \tag{2.119}$$

with $L_{i\ell}$ defined by (2.117). The term $G_{\ell j}(\boldsymbol{r}, t)$ is referred to as transient Green's tensor to be found in the sequel. We however first obtain the response to a point harmonic source.

2.12.1 Steady-State Green's Tensor

We introduce the four-dimensional Fourier transform $f(\boldsymbol{k}, \omega) = f(k_1, k_2, k_3, \omega)$ of a function $f(\boldsymbol{r}, t)$ defined by

$$f(\boldsymbol{k}, \omega) = \int_{V_4} f(\boldsymbol{r}, t) e^{i(\boldsymbol{kr}+\omega t)} dV_4 \quad ,$$

where $dV_4 = dr\,dt = dx_1 dx_2 dx_3 dt$. (Note the difference from the definition for the one-dimensional case given by (1.95)). Then, the transforms, say, of the first spatial and of the second time derivatives of $f(r,t)$ are given by

$$\int_{V_4} (\partial f/\partial x_q) e^{i(kr+\omega t)} dV_4 = -ik_q f(k,\omega)$$

$$\int_{V_4} (\partial^2 f/\partial t^2) e^{i(kr+\omega t)} dV_4 = -\omega^2 f(k,\omega) \quad .$$

Multiplying (2.119) by $e^{i(kr+\omega t)}$ and integrating over V_4 one gets the algebraic equation

$$L_{i\ell}(k,\omega) G_{\ell j}(k,\omega) = -\delta_{ij} \quad ,$$

with

$$L_{i\ell}(k,\omega) = \rho\omega^2 \delta_{i\ell} - (\lambda+\mu) k_i k_\ell - \mu k^2 \delta_{i\ell} \quad . \tag{2.120}$$

These relations provide the expression for $G_{ij}(k,\omega)$,

$$G_{ij}(k,\omega) = \frac{1}{\mu k^2 - \rho\omega^2} \left[\delta_{ij} - \frac{(\lambda+\mu) k_i k_j}{(\lambda+2\mu) k^2 - \rho\omega^2} \right] \quad ,$$

which can be rewritten in terms of the wave velocities, c_α and c_β, as follows:

$$G_{ij}(k,\omega) = \frac{1}{\rho(c_\beta^2 k^2 - \omega^2)} \left[\delta_{ij} - \frac{(c_\alpha^2 - c_\beta^2) k_i k_j}{c_\alpha^2 k^2 - \omega^2} \right] \quad . \tag{2.121}$$

The inversion of this relation yields $G_{ij}(r,t)$

$$G_{ij}(r,t) = (16\pi^4)^{-1} \int_{W_4} G_{ij}(k,\omega) e^{-i(kr+\omega t)} dW_4 \quad , \tag{2.122}$$

where $dW_4 = dk_1 dk_2 dk_3 d\omega$. It is convenient to begin with the inversion with respect to k only, which will provide the so-called steady-state Green's tensor or the response to a point harmonic source.

Thus

$$G_{ij}(r,\omega) = \int_{-\infty}^{\infty} G_{ij}(k,\omega) e^{-ikr} dk/(8\pi^3) \quad ,$$

where $dk = dk_1\, dk_2\, dk_3$. On substituting (2.121) for $G_{ij}(k,\omega)$ the above relation is found to be

$$G_{ij}(r,\omega) = (I_o \delta_{ij} - \partial^2 I_1/\partial x_i \partial x_j)/\rho \quad , \tag{2.123}$$

where

$$I_o = \int_{-\infty}^{\infty} e^{-ikr} dk / \left[(c_\beta^2 k^2 - \omega^2)(8\pi^3) \right] \tag{2.124}$$

$$I_1 = (c_\beta^2 - c_\alpha^2) \int_{-\infty}^{\infty} e^{-ikr} dk / \left[(c_\alpha^2 k^2 - \omega^2)(c_\beta^2 k^2 - \omega^2) 8\pi^3 \right] \quad . \tag{2.125}$$

Evaluation of these integrals (see the comments on Problem 2-17) shows that

$$I_o = e^{i\omega r/c_\beta}/(4\pi r c_\beta^2) \tag{2.126}$$

$$I_1 = (e^{i\omega r/c_\alpha} - e^{i\omega r/c_\beta})/(4\pi r \omega^2) \tag{2.127}$$

with $r = |\boldsymbol{r}|$. The steady-state Green's tensor, $G_{ij}(\boldsymbol{r},\omega)$, is thus given by (2.123), (2.126) and (2.127). It can be generalized to the following convenient form:

$$G_{ij}(\boldsymbol{r},\omega) = \frac{1}{4\pi\rho\omega^2}\left[\beta^2 \frac{e^{i\beta R}}{R}\delta_{ij} - \frac{\partial}{\partial x_i}\frac{\partial}{\partial x_j}\left(\frac{e^{i\alpha R} - e^{i\beta R}}{R}\right)\right] \quad , \tag{2.128}$$

where $\alpha = \omega/c_\alpha$, $\beta = \omega/c_\beta$ and $R = |\boldsymbol{r} - \boldsymbol{r}'|$, with $\boldsymbol{r}'$ indicating the point of application of the source. Note that the reciprocity principle, mentioned in the previous section, takes place again.

2.12.2 Transient Green's Tensor

It remains to carry out the inversion with respect to ω. To this end we, first, deduce a more convenient representation for $G_{ij}(\boldsymbol{r},\omega)$. Let us introduce a unit vector $\nu = (\nu_1,\ \nu_2,\ \nu_3)$ as follows:

$$\nu = \boldsymbol{r}/r \quad .$$

Then

$$r_{,i} = \nu_i, \qquad r_{,ij} = (\delta_{ij} - \nu_i\nu_j)/r, \qquad \nu_i\nu_j = x_i x_j/r^2 \quad .$$

Taking the derivatives in (2.123) one gets

$$G_{ij}(\boldsymbol{r},\omega) = [h(\omega r)\delta_{ij} + g(\omega r)\nu_i\nu_j]/r \tag{2.129}$$

where

$$h(\omega r) = \frac{1}{4\pi\rho\omega^2 r^2}\left\{\left[(1 - \frac{ir\omega}{c})e^{i\omega r/c}\right]\Big|_{c_\beta}^{c_\alpha} + \frac{r^2\omega^2}{c_\beta^2}e^{i\omega r/c_\beta}\right\}$$

$$g(\omega r) = -\frac{1}{4\pi\rho\omega^2 r^2}\left\{\left[3(1 - \frac{i\omega r}{c}) - \frac{r^2\omega^2}{c^2}\right]e^{i\omega r/c}\right\}\Big|_{c_\beta}^{c_\alpha}$$

with

$$f(c)\Big|_{c_\beta}^{c_\alpha} = f(c_\alpha) - f(c_\beta) \quad .$$

The inversion with respect to ω is defined by

$$G_{ij}(\boldsymbol{r},t) = \frac{1}{2\pi}\int_{-\infty}^{\infty} G_{ij}(\boldsymbol{r},\omega)e^{-i\omega t}d\omega \quad ,$$

which differs from (1.94) by a multiplier. Substituting (2.129) for $G_{ij}(r,\omega)$ and taking into account the well-known Fourier representations for the Dirac function, $\delta(t)$, for the unit step function, $h(t)$, and that

$$(1/r)_{,ij} = (3\nu_i\nu_j - \delta_{ij})/r^3 \quad ,$$

the explicit expression for Green's dynamic tensor is finally found to be

$$4\pi\rho G_{ij}(r,t) = \delta(t - r/c_\beta)\delta_{ij}/(c_\beta^2 r)$$
$$+\left[\delta(t - r/c_\alpha)/c_\alpha^2 - \delta(t - r/c_\beta)/c_\beta^2\right]x_i x_j/r^3 + (1/r)_{,ij}tm \quad , \tag{2.130}$$

where

$$m = 1, \qquad \text{for} \quad r/c_\alpha < t < r/c_\beta$$
$$m = 0, \qquad \text{for} \quad t < r/c_\alpha,\ t > r/c_\beta \quad .$$

This expression contains the disturbances of both, P- and S-modes. Its last term describes the motion taking place between the arrivals of these two modes. As was defined in the beginning of this section, $G_{ij}(r,t)$ represents the displacement along the x_i-direction occurring at point r due to the unit impulse applied at $r = 0$ in the x_j-direction at $t = 0$, while $G_{ij}(r,\omega)$ provides the steady-state response. Having determined $G_{ij}(r,t)$, the elastic radiation due to the source with an arbitrary time or space dependence may be set up via convolution integrals, which explains its major role in linear Elastodynamics.

2.13 Energy Transport in Harmonic Waves

In this section we shall specify the energy flux, introduced in Section 1.14, and discuss related concepts for the case of harmonic waves.

Under dynamic loads the energy E, contained in volume V of an elastic medium, varies in time. Its increase, dE, equals the work done by the tractions applied to the boundary surface S of the volume. Assuming that the tractions cause the displacement du during time dt, we get by means of the conservation of energy

$$dE = \int_S\int \sigma_{ik}n_k du_i ds \quad ,$$

where n is the unit normal to the surface element ds. The variation per unit time is, thus, given by

$$\partial E/\partial t = \int_S\int \sigma_{ik}n_k \dot{u}_i ds \quad . \tag{2.131}$$

Defining the Poynting vector, P, by the relation

$$P_k = -\sigma_{ik}\dot{u}_i$$

we get

$$\partial E/\partial t = -\int_S \int P_k n_k ds \quad , \tag{2.132}$$

in accord with the results of Section 1.14.

Thus, the energy variation within volume V per unit time is represented as the flux of the Poynting vector over the boundary surface. The vector $\boldsymbol{P}$ indicates both the rate and the direction of the energy transport.

The defined values are obviously functions of time. To characterise the energy transport by means of a number one takes the average of (2.131) over a typical time interval T. This value, $\langle I \rangle$ is referred to as the average power flux or the energy transfer rate

$$\langle I \rangle = -(1/T)\int_0^T \dot{E}dt = (1/T)\int_0^T \int_S \int P_k n_k ds dt \quad . \tag{2.133}$$

It would be useful to specify the above relation to the case of harmonic waves,

$$\sigma_{ik} = \sigma_{ik}^o(r)e^{-i\omega t}$$
$$u_i = u_i^o(r)e^{-i\omega t} \quad ,$$

where σ_{ik}^o, u_i^o are complex amplitude functions. Of course, only real or imaginary parts of σ_{ik} and u_i should be involved in calculations of the energy. Taking into account that for any complex value Q

$$\mathrm{Re}\,[Q] = (Q + Q^*)/2 \quad ,$$

where asterisk indicates the complex conjugate, we get, in place of (2.133),

$$\langle I \rangle = -(1/4T)\int_0^T \int_S \int n_k(\sigma_{ik}\dot{u}_i + \sigma_{ik}\dot{u}_i^* + \sigma_{ik}^*\dot{u}_i + \sigma_{ik}^*\dot{u}_i^*)dsdt \quad .$$

Since $T = 2\pi/\omega$ and, hence,

$$\int_0^T e^{\mp i2\omega t}dt = 0$$

we get

$$\langle I \rangle = -(1/4)i\omega \int_S \int n_k(\sigma_{ik}^o u_i^{o*} - \sigma_{ik}^{o*} u_i^o)ds \quad . \tag{2.134}$$

Let us compute $\langle I \rangle$ for the case of SH-waves emitted by vibrations of the buried rigid cylinder investigated in Section 2.8. Among the stresses only σ_{rz} is non-zero and is given by

$$\sigma_{rz} = \mu \partial u_z/\partial r = \mu A_o k_\beta^3 H_{-1}^{(1)}(k_\beta r)e^{-i\omega t} \quad .$$

Prescribing $r = R$ as a reference surface and substituting the expressions for σ_{rz} and u_z into (2.134) we get

$$\langle I\rangle = -(1/4)i\omega \int_0^{2\pi} (\sigma_{rz}^o u_z^{o*} - \sigma_{rz}^{o*} u_z^o) \Big|_{r=R} R d\theta$$

$$= -(1/4)i\omega\mu \int_0^{2\pi} \Big[A_o k_\beta^5 H_{-1}^{(1)}(k_\beta R) A_o^* H_o^{(2)}(k_\beta R)$$

$$- A_o^* k_\beta^5 H_{-1}^{(2)}(k_\beta R) A_o H_o^{(1)}(k_\beta R)\Big] R d\theta \quad ,$$

where the relation

$$H_n^{(2)}(k_\beta r) = H_n^{(1)*}(k_\beta r)$$

has been used. This reduces to

$$\langle I\rangle = -(1/2)i\omega\mu |A_o|^2 k_\beta^5 \Big[H_{-1}^{(1)}(k_\beta R) H_o^{(2)}(k_\beta R)$$

$$-H_{-1}^{(2)}(k_\beta R) H_o^{(1)}(k_\beta R)\Big] R\pi \quad . \tag{2.135}$$

Since the medium is perfectly elastic no loss of energy occurs. Taking advantage of this, we put $R\to\infty$ and make use of asymptotic expansions for Bessel functions given by (2.49) to obtain

$$\langle I\rangle = -(1/4)i\omega\mu |A_o|^2 k_\beta^5 (2/\pi k_\beta R) 2iR2\pi = 2\mu\omega |A_o|^2 k_\beta^4 \quad .$$

On the other hand, we could avoid the above limiting procedure by noting that the bracketed expression in (2.135) is the Wronskian

$$-W\{H_o^{(1)}(z),\ H_o^{(2)}(z)\} = 4i/(\pi z) \quad ,$$

which would immediately provide the same results. Employing the expression for A_o given in Section 2.8, we finally find that the average power flux is given by

$$\langle I\rangle = 2\mu\omega (u_c^o)^2 / \Big[J_o^2(k_\beta a) + Y_o^2(k_\beta a)\Big] \quad ,$$

where u_c^o is the amplitude of vibrations of the cylinder and $k_\beta = \omega/c_\beta$.

The strong dependence of the power on frequency is observed from this relation. In fact, for $\omega\to\infty$ we have by (2.49)

$$\langle I\rangle \simeq (u_c^o)^2 \mu\pi\omega k_\beta a \to \infty \quad .$$

$$(k_\beta a) \gg 1 \quad .$$

On the other hand, the expansion for low frequencies is obtained via the following relation:

$$H_o^{(2)}(z) \simeq -2i \ln z/(\pi) \quad ,$$

$$z \ll 1 \quad ,$$

which yields

$$\langle I\rangle \simeq (u_c^o)^2 \mu\pi^2\omega/2\ln^2(k_\beta a)\to 0 \quad ,$$
$$(k_\beta a) \ll 1 \quad .$$

Another value of interest is the intensity, which describes the energy transfer rate per unit area in the direction of propagation, m. We shall denote it as e,

$$e = -\sigma_{ij}\dot{u}_j m_i = P_i m_i \quad , \tag{2.136}$$

and its average value as $\langle e\rangle$. For, say, a plane logitudinal wave of displacement

$$u_i = u^o e^{i(k_\alpha x-\omega t)}$$

the average intensity is

$$\langle e\rangle = (\lambda+2\mu)(u^o)^2\omega^2/(2c_\alpha) \quad . \tag{2.137}$$

The defined values will be of an extensive use in characterizing wave-obstacle interactions and the response of inhomogeneous solids. Section 3.5 deals with extensions of these concepts to the case of anisotropic media.

2.14 Viscoelastic Waves. Spatial and Temporal Attenuation

The correspondence principle, described in Section 1.8, applies in an especially simple way to harmonic viscoelastic waves. In fact, suffice to replace the real wave numbers, k_α and k_β, by their complex generalizations, $k_\alpha(i\omega)$ and $k_\beta(i\omega)$, to make a transition from a purely elastic problem to the associated viscoelastic one. This operation holds under the following limitations: i) there are no boundary conditions explicitly depending on time, and ii) the elastic solution has been obtained without any separation into imaginary and real parts or the determination of the modulus.

Consider, for example, a plane P-wave travelling in an elastic medium and given by

$$u_1 = u_1^o e^{i(k_\alpha x-\omega t)} \quad ,$$

with

$$k_\alpha = \omega/c_\alpha = \omega\rho^{1/2}/(\lambda+2\mu)^{1/2} \quad .$$

The associated wave in a viscoelastic body with the complex moduli $\lambda(i\omega) = \lambda^r(\omega)+i\lambda^i(\omega)$ and $\mu(i\omega) = \mu^r(\omega)+i\mu^i(\omega)$ is

$$u_1 = u_1^o e^{i[k_\alpha(i\omega)x-\omega t]} \quad , \tag{2.138}$$

with

$$k_\alpha(i\omega) = \omega\rho^{1/2}/[\lambda(i\omega)+2\mu(i\omega)]^{1/2} \quad .$$

It is convenient to introduce the complex compliance,

$$S_\alpha(i\omega) = [\lambda(i\omega)+2\mu(i\omega)]^{-1} = S_{r\alpha}(\omega) - iS_{i\alpha}(\omega)$$

or in polar coordinates

$$S_\alpha(i\omega) = |S_\alpha| e^{-i\phi_\alpha} \quad .$$

Then

$$k_\alpha(i\omega) = \omega\sqrt{\rho S_\alpha(i\omega)} = \omega\sqrt{\rho|S_\alpha|}e^{-i\phi_\alpha/2} \quad .$$

Substituting this relation for $k_\alpha(i\omega)$ into (2.138) yields the viscoelastic wave,

$$u_1 = u_1^o e^{-k_\alpha^i(\omega)x} e^{i[k_\alpha^r(\omega)x - \omega t]} \quad ,$$

where

$$k_\alpha^r(\omega) = \mathrm{Re}\,[k_\alpha(i\omega)] = \omega\sqrt{\rho|S_\alpha|}\cos(\phi_\alpha/2)$$
$$k_\alpha^i(\omega) = \mathrm{Im}\,[k_\alpha(i\omega)] = -\omega\sqrt{\rho|S_\alpha|}\sin(\phi_\alpha/2) \quad .$$

Similar expressions hold for shear waves too.

It is seen that the plane viscoelastic waves decay ($x > 0,\ k_\alpha^i(\omega) > 0$) during their propagation. The decay rate is frequency dependent and governed by the imaginary part of the wave number. The wave velocity is given by

$$c_\alpha(\omega) = \omega/k_\alpha^r(\omega) = \left[\rho^{1/2}|S_\alpha|^{1/2}\cos(\phi_\alpha/2)\right]^{-1} \quad .$$

Thus, the velocity is frequency dependent too, and governed by the real part of the complex wave number.

These two effects, which are referred to as the wave attenuation and the wave dispersion, constitute a radical difference between behavior of a purely elastic infinite medium and that of a viscoelastic one. To appreciate the physical consequences it is useful to imagine the propagation of a pulse in each of the media. In a perfectly elastic material the pulse does not suffer any distortion. On the contrary, in a viscoelastic material it decays, due to the attenuation. The pulse is also smeared out, since the harmonics it consists of propagate with different velocities.

In a similar way this particular version of the correspondence principle applies to any of the foregoing problems dealing with harmonic waves, including Green's steady-state tensor. One concludes that in a viscoelastic isotropic material there are dilatational as well as transverse shear waves traveling with the frequency-dependent velocities and the attenuation coefficients.

Real and imaginary parts of the complex wave number, $k(i\omega)$, are not independent of one another. In fact, the function

$$H(i\omega, x) = H^r(\omega, x) + iH^i(\omega, x) = e^{ik(i\omega)x}$$

may be viewed as the admittance of a viscoelastic medium, since it relates between the disturbance at the boundary $x = 0$ (input) and that at the distance x (output). This may imply the Kramers-Kronig ($K - K$) relations given by (1.108), which impose a dependence between $\mathrm{Re}\, H(i\omega, x)$ and $\mathrm{Im}\, H(i\omega, x)$.

The logarithm of an analytic function is analytic in the same region, except at the zeros. On assuming that the medium is not a perfect reflector, to exclude the occurrence of zeros, we may conclude that $k(i\omega)$ is also analytic in the upper half-plane. Under some additional conditions of regularity, which can be found in literature, the following K-K relation may be deduced with the help of (1.108):

$$n(i\omega)-n(i\omega_o) = \left(\frac{\omega-\omega_o}{\pi i}\right) P \int_{-\infty}^{\infty} \frac{n(i\omega')d\omega'}{(\omega'-\omega)(\omega'-\omega_o)} \quad , \tag{2.139}$$

where $n(i\omega) = \tilde{c}k(i\omega)/\omega$, ω_o is an arbitrary reference frequency, which may be infinity or zero in particular cases, and $\tilde{c}$ reference velocity. To insure that a real disturbance remains real after travelling through the medium, the entity, $n(i\omega)$, must satisfy the identity

$$n(-\omega) = n^*(\omega) \quad ,$$

with asterisk denoting complex conjugate. Then separation of (2.139) into real and imaginary parts provides two equations for $\mathrm{Re}\, n(i\omega)$ and $\mathrm{Im}\, n(i\omega)$. For example,

$$\begin{aligned} \mathrm{Re}\, n(i\omega) &= \mathrm{Re}\, n(0) + \frac{2\omega^2}{\pi} P \int_0^{\infty} \frac{\mathrm{Im}\, n(i\omega')d\omega'}{\omega'(\omega'^2-\omega^2)} \\ \mathrm{Re}\, n(i\omega) &= \mathrm{Re}\, n(\infty) + 2P \int_0^{\infty} \frac{\mathrm{Im}\, n(i\omega')\omega' d\omega'}{\pi(\omega'^2-\omega^2)} \quad , \end{aligned} \tag{2.140}$$

which hold if

$$\lim_{\omega \to \infty} \mid n(\omega)/\omega \mid = 0.$$

Now we go over to descriptions of the attenuation. The amplitude of a plane viscoelastic wave propagating, say, in a half-space, $x \geq 0$, can be written as

$$A(x) = A(0)e^{-\alpha x} \quad ,$$

with $\alpha = k^i(\omega)$ the attenuation (absorption) coefficient given by

$$\alpha = (1/x)\ell n[A(0)/A(x)] \quad .$$

In this expression α has dimensions of reciprocal length, and is usually denoted as Np/m (nepers per meter). On employing decibels one can write

$$\alpha(dB/m) = (20/x)\log_{10}[A(0)/A(x)] = 8.686\alpha(Np/m) \quad .$$

The intensity, e, and other energy values are proportional to the square of amplitude, which implies

$$e(x) = e(0)e^{-2\alpha x} \tag{2.141}$$

and the energy loss is governed by

$$de/dx = -2\alpha e. \tag{2.142}$$

Thus, during the propagation through a distance of one wavelength, λ, the relative energy loss is $2\alpha\lambda$, which, in view of (1.48), yields

$$\alpha = \delta/\lambda \quad ,$$

with δ the logarithmic decrement.

In the case of harmonic vibrations, on the other hand, the attenuation β describes the temporal decay of a disturbance for constant x

$$B(t) \simeq e^{-\beta t} \quad ,$$

with $\beta = \delta\omega/(2\pi)$. Thus, the coefficients of temporal and spatial decays, β and α, respectively, and the wave velocity c, are related by

$$\beta/\alpha = c \quad , \tag{2.143}$$

which is useful for experimental characterization of attenuative media.

2.15 Viscoelastic Pulse Propagation. Wavefront Velocity

Specific features of transient radiation in a viscoelastic medium may be conveniently shown on an example of a spherical cavity, which is suddenly subject to a pressure of magnitude p_o decreasing rapidly thereafter. Besides the obvious practical interest, this problem enables one to draw quite general conclusions on transient motions in lossy materials.

The impact loading described above can be approximated by imposing the boundary conditions

$$\sigma_{rr}(a,t) = -p_o\delta(t)$$

with a denoting radius of the sphere.

A better understanding of the physics of this problem may be achieved by considering firstly harmonic pulsations of the cavity. The solution for this case, σ^h_{rr}, follows from (2.110) and (2.111) and the viscoelastic analogy

$$\sigma^h_{rr}/p_o = -H_V(r,i\omega)e^{-i\omega t} \quad ,$$

with the viscoelastic admittance function given by

$$H_V(r,i\omega) = \frac{a^3\{\omega^2\rho r^2 + 4\mu(i\omega)[ik_\alpha(i\omega)r - 1]\}}{r^3\{\omega^2\rho a^2 + 4\mu(i\omega)[ik_\alpha(i\omega)a - 1]\}} e^{ik_\alpha(i\omega)(r-a)} \quad . \tag{2.144}$$

Since the Fourier transform of $\delta(t)$ is $(\sqrt{2\pi})^{-1}$ (see (1.95)) and the system is linear, the transient stress, σ_{rr}, is given by

$$-\sigma_{rr}(r,t)/p_o = (2\pi)^{-1}\int_{-\infty}^{\infty} H_V(r,i\omega)e^{-i\omega t}d\omega \quad , \tag{2.145}$$

similarly to (2.114).

The evaluation of this integral may be carried out in the complex ω-plane. For $\tau = [t - (r-a)/c(\omega)] < 0$ the causality constraint requires

$$\sigma_{rr}(r,\tau) = 0, \quad \tau < 0 \quad . \tag{2.146}$$

This is appropriate to integration in the upper half-plane due to the location of the roots of the denominator in (2.144), which lie in the lower half-plane.

For $\tau > 0$ the contour lies in the lower half-plane and the stress under proper regularity conditions is given by

$$\sigma_{rr}(r,t) = -2\pi i \sum res\left[(2\pi)^{-1} H_V(r, i\omega) e^{-i\omega t}\right] p_o \tag{2.147}$$

$(\tau > 0)$

On the understanding that (2.146) holds as $\omega \to \infty$ we observe that a point $r \gg a$ first learns about the pulse at the instant $t = (r-a)/c_\alpha(\infty)$. Hence a wave front propagates with the speed

$$c_\infty = \lim_{\omega \to \infty} c(\omega) \tag{2.148}$$

in agreement with the concept of signal velocity considered in Section 1.16.

The above remarks rely on physical arguments rather than on a rigorous mathematical analysis. Nevertheless, they make possible to draw a general conclusion on the front speed in dispersive media, without specifying a particular viscoelastic model. This conclusion holds for other types of sources too.

Details of the progressive pulse may become clearer, if the moduli, $\lambda(i\omega)$ and $\mu(i\omega)$, of the material are known, so the integral may be evaluated. The exact inversion may, however, turn out to be a tedious procedure.

For a particular case of a purely elastic medium the inverse is available in tables. This provides the displacement, u_r, which may be conveniently written as

$$u_r(r,t) = -\frac{a^3 p_o}{4\mu} \frac{d}{d\tau} \frac{d}{dr} \left\{1 - (2-2\nu)^{1/2} e^{\omega_i \tau} \sin\left[\omega_r \tau + \cot^{-1}(1-2\nu)^{(1/2)}\right]\right\} \frac{h(\tau)}{r} , \tag{2.149}$$

where

$$\pi/4 \le \cot^{-1}(1-2\nu)^{1/2} \le \pi/2$$

$$\omega_i = \operatorname{Im} \omega_{res}$$

$$\omega_r = |\operatorname{Re} \omega_{res}|$$

$$\tau = t - (r-a)/c_\alpha$$

and $h(\tau)$ is a step function. It is seen that the real part of the complex resonance frequency, ω_{res}, given by (2.113) is indeed responsible for the "periodicity", and its imaginary part for the temporal decay, as noted in Section 2.11.

2.16 Radiation from a Moving Dislocation

Fracture and plastic flow are believed to be associated with motions of dislocations. These also contribute to elastic noises available in solids. Investigations of radiation due to simple models of microdefects may therefore provide a useful insight in the above phenomena, at least from a qualitative point of view. Consider a screw dislocation located in an infinitely extended medium (Figure 1-7). This dislocation constitutes an alternating slip across the x_1-axis concentrated at the origin and has the Burgers vector $\boldsymbol{b} = (0, 0, b)$. The static displacement along the x_3-axis (coinciding with the dislocation line AB) is given by (see Problem 1-5).

$$u_3^s = b\theta/(2\pi) = b\tan^{-1}(x_2/x_1)/(2\pi) \quad . \tag{2.150}$$

Now imagine that the AB-line executes vibrations along the x_1-axis. Thus, a position of the dislocation center is

$$x_1^{AB} = de^{-i\omega t}, \qquad x_2^{AB} = 0 \quad . \tag{2.151}$$

The quasistatic displacement, u_3^{qs}, corresponding to this motion of the dislocation core, follows in a straightforward manner as an extension of (2.150)

$$u_3^{qs} = \frac{b}{2\pi}\tan^{-1}\left(\frac{x_2}{x_1 - de^{i\omega t}}\right) = \frac{b}{2\pi}\theta + \frac{b}{2\pi}\frac{\sin\theta}{r}de^{-i\omega t} \quad . \tag{2.152}$$

However, this function is not a solution to the wave equation and can therefore only be used in the vicinity of the source, when $r \to 0$. Thus, (2.152) provides the following restriction on the displacement u_3:

$$u_3 \simeq u_3^{qs}, \qquad \text{as} \quad r \to 0 \quad . \tag{2.153}$$

It is convenient to represent u_3 as a sum of the static solution u_3^s, given by (2.150), and the dynamic one, u_3^d

$$u_3 = u_3^s + u_3^d \quad . \tag{2.154}$$

Then the problem is reduced to finding u_3^d, which, according to (2.150) - (2.154), is to satisfy the relation

$$\lim_{r \to 0} u_3^d = \frac{b}{2\pi}\frac{\sin\theta}{r}de^{-i\omega t} \quad . \tag{2.155}$$

In order to find u_3^d we first examine the simplest case of SH-waves. These constitute the displacement in the x_3-direction which propagates in the normal $r\theta$-plane. Adapting (2.25) one gets

$$\nabla^2 u_3^d = \frac{1}{r}\frac{\partial}{\partial r}\left(r\frac{\partial u_3^d}{\partial r}\right) + \frac{1}{r^2}\frac{\partial^2 u_3^d}{\partial \theta^2} = \frac{1}{c_\beta^2}\ddot{u}_3^d \quad .$$

According to (2.52) and (2.53)

$$u_3^d = \sum_n \left[A_n H_n^{(1)}(k_\beta r) + B_n H_n^{(2)}(k_\beta r)\right] e^{in\theta} e^{-i\omega t} \quad ,$$

where the coefficients B_n should be set to zero, $B_n = 0$, to represent outgoing waves. Invoking now (2.155) and the expansions of the Hankel function,

$$\lim_{z \ll 1} H_o^{(1)}(z) \simeq iY_o(z) = \frac{2i}{\pi}\ell n z$$

$$\lim_{z \ll 1} H_n^{(1)}(z) \simeq iY_n(z) = -\frac{i}{\pi}\left(\frac{2}{z}\right)^n (n-1)! \quad (n \geq 1)$$

one finds

$$A_1 = bdk_\beta/4$$

$$A_n = 0 \quad \text{for} \quad n \neq 1 \quad .$$

Thus, the radiated field is given by

$$u_3^d(r,\theta,t) = bdk_\beta H_1^{(1)}(k_\beta r)\sin\theta e^{-i\omega t}/4. \tag{2.156}$$

On the basis of the correspondence principle this solution is generalized to the case of viscoelastic media by replacing the real wave number, k_β, with the complex wave number, according to the adopted viscoelastic model:

$$k_\beta \rightarrow k_\beta(i\omega) = k^r(\omega) + ik^i(\omega) = \omega\sqrt{\rho S(i\omega)} \quad .$$

Clearly, the Hankel function appearing in (2.156) becomes one of a complex argument.

Making use of the Fourier integral, it is possible to show (see comments on Problem 2-22) that the transient radiation due to a step-motion of the dislocation,

$$x_1^{AB} = dh(t) = \begin{cases} d, & t > 0 \\ 0, & t < 0 \end{cases}$$

$$x_2^{AB} = 0$$

is given by an attractively simple expression

$$u_3^d = \frac{bd}{2\pi r}\frac{\sin\theta}{(c_\beta^2 t^2 - r^2)^{1/2}}(c_\beta t)h(c_\beta t - r) \quad . \tag{2.157}$$

Note that (2.156) and (2.157) are, in fact, physically meaningful only in the far-field approximation, $r \rightarrow \infty$, since in the source proximity the radiation is affected by plastic strains due to the singularity in the displacement field.

2.17 Creation of a Dislocation Dipole. Nonuniform Motion

Imperfections birth or their annihilation illustrates further a wide variety of wave sources in solids. Such events, in fact, occur spontaneously or while loading process. Consider the example of a sudden creation of a dislocation dipole, which, besides the usefulness of its own, provides a remarkable solution to a nonuniform motion of dislocation.

We seek for radiation due to rapid transition of a virgin elastic medium at rest, $u_i = 0$, to the state corresponding to the presence of a dislocation dipole. It is supposed that the birth takes place at time $t = \tau$.

Write first the static displacement u_3^s for a single screw dislocation situated at $x_1 = \varepsilon$, $x_2 = 0$. It follows from (2.150)

$$u_3^s = b\tan^{-1}[x_2/(x_1 - \varepsilon)]/2\pi \quad .$$

Adding the dislocation of opposite sign at $x_1 = \varepsilon - \delta\varepsilon$, $x_2 = 0$, $(\delta\varepsilon \to 0)$ we obtain the static infinitesimal field, δu_3^s, of the dipole

$$\delta u_3^s = -\frac{\partial u_3^s}{\partial x_1}\delta\varepsilon = bx_2\delta\varepsilon/(2\pi R^2) \quad , \tag{2.158}$$

with

$$R^2 = (x_1 - \varepsilon)^2 + x_2^2 \quad .$$

For the dynamic effects have been neglected while deriving (2.158), we think of it as though it were a relation describing the field after a long time the dipole has been produced. In other words, if radiation associated with the dipole birth at $t = \tau$ is denoted as δu_3^b, then

$$\delta u_3^b \to bx_2\delta\varepsilon/(2\pi R^2) \quad \text{as} \quad t \to \infty \quad . \tag{2.159}$$

On the other hand, δu_3^b must clearly satisfy the initial condition

$$\delta u_3^b \to 0 \quad \text{as} \quad t \to \tau \tag{2.160}$$

and the equation of motion for SH-waves

$$u_{3,11} + u_{3,22} = \ddot{u}_3/c_\beta^2 \quad . \tag{2.161}$$

To this end, consider the function

$$\delta u_3^b = bx_2\delta\varepsilon f(q)h[c_\beta(t-\tau) - R]/(2\pi R^2) \quad , \tag{2.162}$$

where $q = R/[c_\beta(t-\tau)]$. Clearly, the mere role of the step function, $h(\xi)$, is to insure the causality of radiation, which should vanish when $R > c_\beta(t-\tau)$. The unknown yet function, $f(q)$, may be specified from (2.159) – (2.161).

Indeed, substitution of (2.162) into (2.161) yields

$$q(1-q^2)f''(q) - (2q^2+1)f'(q) = 0 \quad ,$$

which has the following solution satisfying (2.160)

$$f(q) = (1/q^2 - 1)^{-1/2}/q = c_\beta(t-\tau)\left[c_\beta^2(t-\tau)^2 - R^2\right]^{-1/2} \quad .$$

Accordingly, the dynamic displacement field of a dipole created at time $t = \tau$ is eventually given by

$$\delta u_3^b/\delta\varepsilon = bx_2c_\beta(t-\tau)\left[c_\beta^2(t-\tau)^2 - R^2\right]^{-1/2} \\ h[c_\beta(t-\tau) - R]/(2\pi R^2) \quad . \tag{2.163}$$

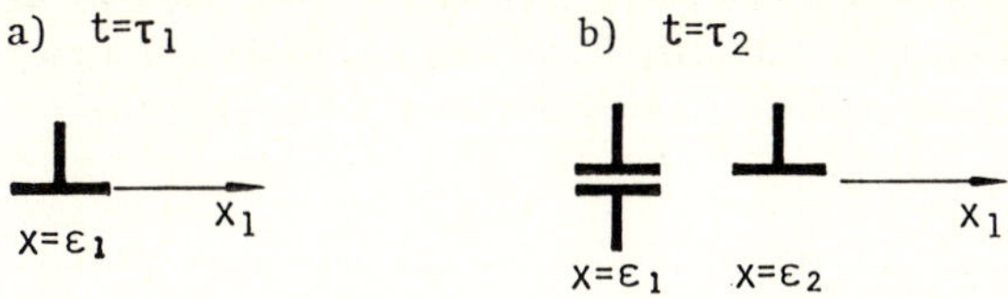

Fig. 2-9 Modelling a dislocation motion by instantaneous creations of dipoles.

This expression provides a remarkable way of deducing the radiation from an arbitrarily moving dislocation. Figure 2-9 illustrates the idea. At time $t = \tau_1$ there is a stationary single dislocation located, say, at $x = \varepsilon_1$, Figure 2-9a. Then a dipole is created at $t = \tau_2 = \tau_1 + \delta\tau$ so as to "move" this dislocation to a new position $x = \varepsilon_2 = \varepsilon_1 + \delta\varepsilon$, and so forth, Figure 2-9b. Indeed, two dislocations of opposite sign placed together cancel each other. By superposing this jerky motion, the field of an arbitrarily moving imperfection may be constructed. For example, assume that the speed of dislocation motion along the x_1-direction is given by $v(t)$. Then summing the "jumps" described by (2.163) we get the radiated field

$$u_3^b(x_1, x_2, t) = \frac{bc_\beta}{2\pi} \int_{-\infty}^{t_o} \frac{(t-\tau)x_2 v(\tau)d\tau}{R^2[c_\beta^2(t-\tau)^2 - R^2]^{1/2}} \quad , \tag{2.164}$$

where use has been made of the relation $\delta\varepsilon = v(\tau)d\tau$ and t_o is the root of the equation

$$x_1^2 + x_2^2 - c_\beta^2(t - t_o)^2 = 0 \quad ,$$

which is less than t. The motion is supposed to start from $x_1 = x_2 = 0$.

Equation (2.164) enables us to compute radiation from an arbitrarily moving dislocation provided that its velocity, $v(t)$, is known. The same technique applies to a plane motion of dislocations.

2.18 Limitations of the Linear Theory. Shock Waves

The foregoing analysis becomes incorrect in the case of intensive and rapid loading, which may cause plastic strains or large deformations. Then, for example, the method of Fourier integrals is no longer valid, since the superposition principle does not apply to non-linear systems. In this sense, the solutions obtained above are of limited use. The exact formulation of the conditions, under which the linearized theory holds, requires a treatment of the non-linear theory. For this the reader should consult the references given at the end of this Chapter.

Fortunately, many relevant problems of a non-linear response may be solved in a reasonably accurate and simple way by employing the concept of a moving discontinuity, given in Sections 1.11 and 1.12. To show some of the new phenomena, which are due to a non-linear behavior, and illustrate elementary related ideas, consider a homogeneous solid, which has an arbitrary cross-section and is semi-infinite in the axial direction. Hence, a bar, a plate, or a half-space may be an example, Figure 2-10. Assume that the left end of this body is given a step velocity, v, which causes the wave front to run towards the right with velocity c.

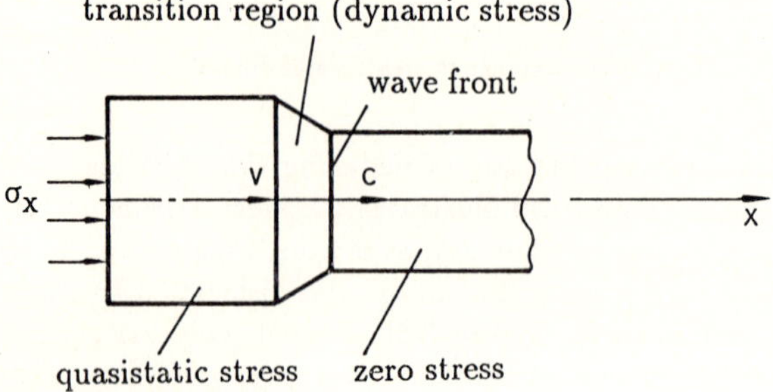

Fig. 2-10 Shock wave running along a bar.

After a short transition time the running wave front should reach steady state motion and separate the stressed region from the undisturbed one. We assume the presence of three spatial domains, as Figure 2-10 illustrates. First, to the right of the wave front there are zero stresses and zero particle velocity. Second, in a small region immediately behind the front a complicated dynamic field may exist. And, finally, to the left of this region the medium is compressed and is characterized by the constant stress σ_x, and the particle velocity, v. It is seen that if the small transition region in the front tail is neglected, the phenomenon may be viewed as a discontinuity wave, which implies, among others, applicability of (1.72) and (1.73).

We assume that the shock can be approximated mathematically as a step-function excitation. However, in reality there is always a finite rise time, and to take this into account, we imagine the excitation to consist of a number of small but finite wavelets, as shown in Figure 2-11. Assume that at, say, the third wavelet the induced stress equals the yield limit k_y. (The constitutive law is pictured in Figure 2-12).

If each step is small enough to cause only negligible geometric changes, we get the estimates that the first three wavelets run at the speed

$$c_1^2 = E_1/\rho \quad , \tag{2.165}$$

while the last two at the speed

$$c_2^2 = E_2/\rho \quad . \tag{2.166}$$

However, if $E_2 < E_1$, as shown in Figure 2-12, then $c_1 > c_2$. Moreover, (2.166) needs the following correction. The faster wave should disturb the region ahead of the slower one. If the particle velocity behind the first wave front is v then the speed of the second is $c_2 + v$ (relatively to the undisturbed material).

Clearly, the two groups of waves occur provided that the stress reaches the yield limit. The wavelets 1-3 constitute the faster wave front, whereas 4-5 the

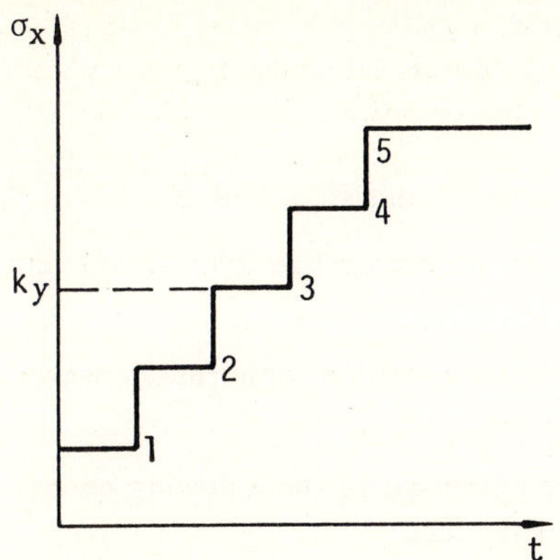

Fig. 2-11 Modelling the step-function load.

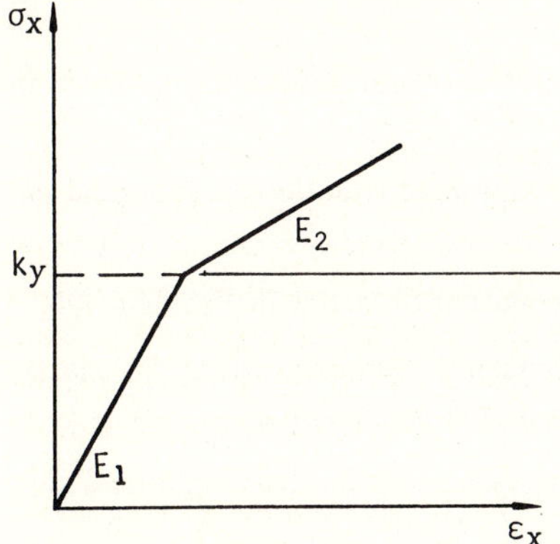

Fig. 2-12 Stress vs. strain.

slower one. It is of interest to interpret these results in terms of the elastic-plastic behavior. They show that an elastic precursor wave will outstrip a plastic wave. Also, in the case of an elastic-perfectly-plastic material, when $E_2 = 0$, no plastic wave may propagate under the above conditions of a uniaxial stress.

Materials with the stress-strain dependence concave upwards, exhibit completely different behavior. Applying the same approach we find that, this time, the group consisting of the low stress wavelets travels slower than the high stress group. Hence, the latter will soon catch up with the low stress wavelets and the single wave front may be formed.

In conclusion, if a material has a "soft" stress-strain relation, concave downward, impact of compression will cause the propagating wave spread out, and consequently, no consistent wave front may result. On the other hand, for materials with a "rigid" stress-strain dependence, concave upward, the shock forms the single wave of discontinuity.

Problems

2-1 Prove that $c_\alpha > c_\beta\sqrt{2}$.

2-2 Write down c_α and c_β in terms of K and ν, E and K, μ and E.

2-3 What is the difference between relations governing the dilatational and shear wave potentials?

2-4 Represent the following running waves as superpositions of harmonic waves: i) step-function wave, ii) delta-function wave.

2-5 Derive the stresses in terms of the wave potentials in the following coordinate systems: i) circular cylindrical, ii) spherical.

2-6 Show that (2.55) represents standing (non-propagating) waves.

2-7 A running wave of displacement does not depend on and is symmetric about the z-axis. Derive the governing differential equation in circular cylindrical coordinates.

2-8 Derive the equation governing a spherically-symmetric running wave in spherical coordinates.

2-9 Check if a function $f(ct\pm r)/r$ satisfies the wave equation in spherical coordinates.

2-10 Recover (2.88) from a solution to the equation obtained in Problem 2-8.

2-11 Derive the low-frequency and high frequency approximations for elastic radiation from a cylindrical cavity, considered in Section 2.7.

2-12 Derive the phase shift for elastic steady-state radiation from spherical cavities.

2-13 Write down the real and imaginary parts of the admittance function for the radiation due to torsional oscillations of an embedded rigid sphere.

2-14 Deduce (2.79) in a simpler way.

2-15 Derive Green's static tensor, given by

$$G_{ij}(r) = (8\pi\mu)^{-1}[r_{,pp}\delta_{ij} - (\lambda+\mu)r_{,ij}/(\lambda+2\mu)]$$

from (2.130).

2-16 Is the relation, $G_{ij}(r,t) = G_{ji}(r,t)$, true?

2-17 Evaluate the integrals given by (2.124) and (2.125) by the residue theorem.

2-18 Imagine two equal and opposite forces acting in the direction of x_1-axis (Figure 2-13). When $h\to 0$ they create a double force without moment. Show that for the unit impulse the disturbance is given by

$$u_i(r,t) \approx \partial G_{i1}(r,t)/\partial x_1$$

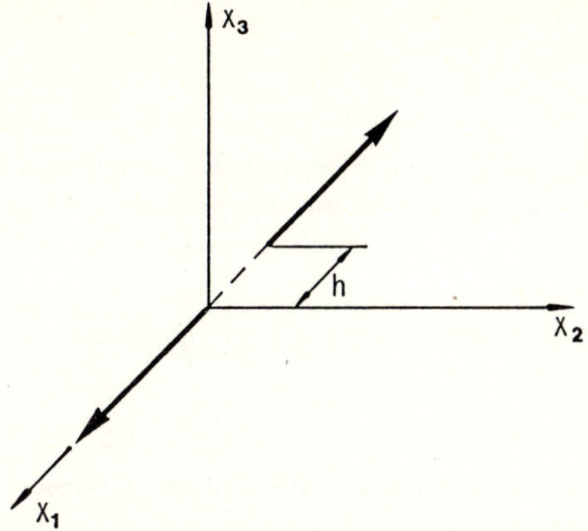

Fig. 2-13 Double force without moment.

2-19 Derive the wave due to the compression center, shown in Figure 2-14 by means of superposition of double forces without moment described in Problem 2-19.

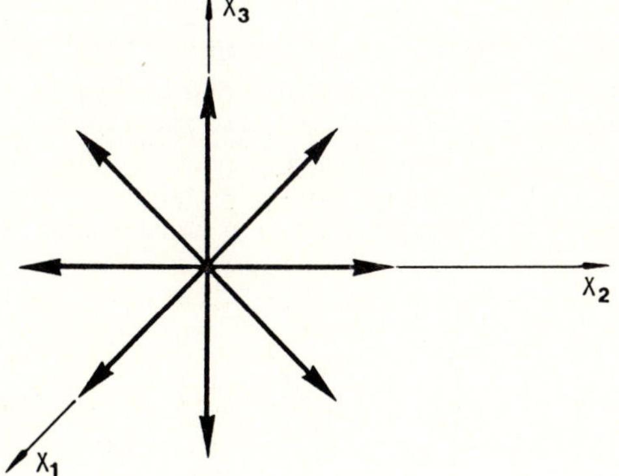

Fig. 2-14 Compression center.

2-20 Derive the average power flux of the pulsating sphere considered in Section 2.9. Derive the acoustic power of a monopole as a low-frequency expansion.

2-21 Why do the amplitudes of the cylindrical and spherical waves decay when $r \to \infty$?

2-22 Confirm (2.157) for radiation due to the step-motion of a screw dislocation.

References and Additional Reading

Sections 2.1 – 2.9

For extensive treatments see Achenbach (1973), Ben-Menachem and Singh (1981), Graff (1975), Miklowitz (1978), Pao and Mow (1973).

Sections 2.10

For a different and extended analysis see Chadwick and Trowbridge (1967).

Section 2.12

Eason, Fulton and Sneddon (1955-1956) provide analysis of Green's dynamic tensors, Miklowitz (1978).

Sections 2.14 – 2.15

Christensen (1971) treats also viscoelastic waves for which the directions of propagation and decay are not identical.

Sections 2.16 – 2.17

Eshelby (1953), Mura (1982), Weertman and Weertman (1980). The ingeneous technique of Section 2.17 is due to Nabarro and to Eshelby.

Section 2.18

A comprehensive treatment of shock waves is given in Eringen and Suhubi (1974-1975).

Chapter III

Bulk Waves in Anisotropic Media

The condition of isotropy does not hold for many natural and man-made materials, such as wood, monocrystals, carbon or graphite fibers, which motivates investigations of the response of anisotropic media. This response is found to be markedly different from that treated in the previous chapter.

A relevant point to note is that phases of a composite material may be isotropic. They are also of finite volumes. Anisotropy of such composites, if any, is therefore of overall nature, unlike point anisotropy of monocrystals.

The fundamental fact of sole existence of pure longitudinal and shear modes discovered for isotropic materials is no longer valid. Only under certain conditions of symmetry such modes may exist in anisotropic media. One of the applications of the presented theory is determination of elastic moduli of anisotropic materials by measuring proper wave speeds.

The piezoelectric waves are of use in many engineering devices. We present the so-called quasi-electrostatic theory of these waves, which combines the dynamic elastic field with the quasi-static electric one and provides a reasonable way of treating piezoelectric phenomena. An approximate analysis of a bulk wave transducer is also given. A related effect of surface piezoelectric waves is considered in the next chapter.

3.1 Plane Waves

Plane elastic waves in an arbitrary anisotropic media are first to be investigated. On neglecting body forces, the equations of motion (1.15) become

$$\sigma_{ij,j} = \rho \ddot{u}_i \quad .$$

Substitution of Hooke's law by

$$\sigma_{ij} = c_{ijk\ell} u_{\ell,k}$$

yields a system of three second order differential equations with respect to the displacement vector,

$$\rho \ddot{u}_i = c_{ijk\ell} u_{\ell,jk} \quad . \tag{3.1}$$

In the case of plane waves we set

$$u_i = A_i f(\tau) \quad , \tag{3.2}$$

where A_i is the amplitude and τ is the local time, given by

$$\tau = t - \boldsymbol{n} \cdot \boldsymbol{r}/V = t - n_j x_j/V \quad ,$$

with $\boldsymbol{r}$ and V the position vector of a particle and the phase velocity, respectively. The propagation direction, $\boldsymbol{n}$, appearing in the relation above, is assumed given. The time and spatial derivatives in (3.1) are as follows:

$$\begin{aligned} \ddot{u}_i &= A_i \partial^2 f/\partial \tau^2 \\ u_{\ell,j} &= -A_\ell n_j (\partial f/\partial t)/V \\ u_{\ell,jk} &= A_\ell n_j n_k (\partial^2 f/\partial \tau^2)/V^2 \quad . \end{aligned}$$

Substituting these expressions into (3.1) yields Christoffel's equation,

$$\rho V^2 A_i = c_{ijk\ell} n_j n_k A_\ell \quad , \tag{3.3}$$

which relates the amplitudes and phase velocities of plane waves, provided the elastic constants, $c_{ijk\ell}$, the density, ρ, and the direction of propagation, $\boldsymbol{n}$, are given.

It is convenient to introduce the second rank tensor, $\Gamma_{i\ell}$, given by

$$\Gamma_{i\ell} = c_{ijk\ell} n_j n_k$$

or, in the explicit expansion with respect to the dummy indices j and k,

$$\begin{aligned} \Gamma_{i\ell} = {} & c_{i11\ell} n_1^2 + c_{i22\ell} n_2^2 + c_{i33\ell} n_3^2 + (c_{i12\ell} + c_{i21\ell}) n_1 n_2 \\ & + (c_{i13\ell} + c_{i31\ell}) n_1 n_3 + (c_{i23\ell} + c_{i32\ell}) n_2 n_3 \quad . \end{aligned} \tag{3.4}$$

Now, (3.3) can be written as

$$(\Gamma_{i\ell} - \rho V^2 \delta_{i\ell}) A_\ell = 0 \quad . \tag{3.5}$$

Expression (3.5) is a set of three homogeneous linear equations with respect to the amplitudes A_ℓ. The solution exists if the determinant vanishes

$$|\Gamma_{i\ell} - \rho V^2 \delta_{i\ell}| = 0 \quad . \tag{3.6}$$

Since the phase velocities, associated with the amplitude A_ℓ and appearing in (3.5), are also unknown, (3.6) applies to specify their values. Being of the third order, this algebraic equation provides three roots for V^2 for any given direction $\boldsymbol{n}$. Knowing V, the amplitudes A_ℓ, or more exactly, the ratio among them, can be determined from (3.5)

The aforementioned remarks outline the procedure. Now consider the technique in greater detail. Due to the symmetry of the tensor, $c_{ijk\ell}$ (see Section 1.7), the tensor $\Gamma_{i\ell}$ is also symmetric. Accordingly, (3.4) provides its components:

$$\begin{aligned}
\Gamma_{11} &= c_{11}n_1^2 + c_{66}n_2^2 + c_{55}n_3^2 + 2c_{16}n_1n_2 + 2c_{15}n_1n_3 + 2c_{56}n_2n_3 \\
\Gamma_{12} &= c_{16}n_1^2 + c_{26}n_2^2 + c_{45}n_3^2 + (c_{12}+c_{66})n_1n_2 \\
&\qquad + (c_{14}+c_{56})n_1n_3 + (c_{46}+c_{25})n_2n_3 \\
\Gamma_{13} &= c_{15}n_1^2 + c_{46}n_2^2 + c_{35}n_3^2 + (c_{14}+c_{56})n_1n_2 \\
&\qquad + (c_{13}+c_{55})n_1n_3 + (c_{36}+c_{45})n_2n_3 \\
\Gamma_{22} &= c_{66}n_1^2 + c_{22}n_2^2 + c_{44}n_3^2 + 2c_{26}n_1n_2 + 2c_{46}n_1n_3 + 2c_{24}n_2n_3 \\
\Gamma_{23} &= c_{56}n_1^2 + c_{24}n_2^2 + c_{34}n_3^2 + (c_{46}+c_{25})n_1n_2 \\
&\qquad + (c_{36}+c_{45})n_1n_3 + (c_{23}+c_{44})n_2n_3 \\
\Gamma_{33} &= c_{55}n_1^2 + c_{44}n_2^2 + c_{33}n_3^2 + 2c_{45}n_1n_2 + 2c_{35}n_1n_3 + 2c_{34}n_2n_3 \\
\Gamma_{21} &= \Gamma_{12}, \quad \Gamma_{31} = \Gamma_{13}, \quad \Gamma_{32} = \Gamma_{23} \quad .
\end{aligned} \tag{3.7}$$

Equation (3.6) then becomes

$$\begin{vmatrix} \Gamma_{11}-\rho V^2 & \Gamma_{12} & \Gamma_{13} \\ \Gamma_{21} & \Gamma_{22}-\rho V^2 & \Gamma_{23} \\ \Gamma_{31} & \Gamma_{32} & \Gamma_{33}-\rho V^2 \end{vmatrix} = 0 \quad . \tag{3.8}$$

The three roots of this cubic equation, V_1^2, V_2^2, and V_3^2, define the phase velocities for a given tensor $\Gamma_{i\ell}$. Now one finds the amplitudes A_ℓ, associated with each of the above velocities, V_j, making use of (3.5), which, in expanded form, is

$$\begin{aligned}
(\Gamma_{11}-\rho V^2)A_1 + \Gamma_{12}A_2 + \Gamma_{13}A_3 &= 0 \\
\Gamma_{21}A_1 + (\Gamma_{22}-\rho V^2)A_2 + \Gamma_{23}A_3 &= 0 \\
\Gamma_{31}A_1 + \Gamma_{32}A_2 + (\Gamma_{33}-\rho V^2)A_3 &= 0 \quad .
\end{aligned} \tag{3.9}$$

Thus, the procedure enables one to find both, the velocity and the displacement vector (polarization), for any given propagation vector $\boldsymbol{n}$.

The important point is that, in general, solutions to (3.8) and (3.9) do not satisfy a priori the relations $\boldsymbol{u} \times \boldsymbol{n} = 0$ or $\boldsymbol{u} \cdot \boldsymbol{n} = 0$, which define purely longitudinal or purely shear waves, respectively, according to Sections 2.1 and 2.2 (Another definition of pure modes will be given in Section 3.5). Thus, for an arbitrary direction there is no pure P- or S-mode in anisotropic media. The situation is schematically depicted in Figure 3-1. The wave whose displacement is closest to $\boldsymbol{n}$(u_{qp} in Figure 3-1) is sometimes referred to as quasi-longitudinal, whereas the two others – as quasi-shear waves. It should be pointed out that the displacements, u_{qp} and u_{qs}, are, by definition, mutually orthogonal, but no displacement is necessarily parallel or normal to $\boldsymbol{n}$. Nevertheless, for materials with symmetry it is possible to find specific directions of propagation for which pure modes do occur. This phenomenon is investigated in the next section.

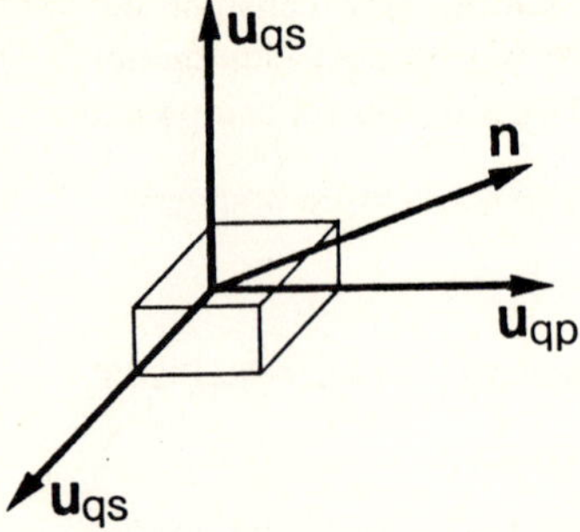

Fig. 3-1 General modes in anisotropic media.

3.2 Effects of Symmetry

Consider the case when the x_1x_2 – plane is that of elastic symmetry (mirror). Either a unidirectional fibrous composite shown in Figure 1-5 or a monoclinic crystal, provides an example of such a medium. One can expect, by symmetry arguments, that propagation along the x_3-axis is associated with a purely longitudinal mode, as well as with two pure shear wave polarized in the x_1x_2-plane (see Figure 1-5).

Let us check this assumption. Setting $n_1 = n_2 = 0$, $n_3 = 1$, and making use of (3.7), one finds that the propagation tensor, $\Gamma_{i\ell}$, is

$$\Gamma_{i\ell}(x_3) = \begin{bmatrix} c_{55} & c_{45} & c_{35} \\ c_{45} & c_{44} & c_{34} \\ c_{35} & c_{34} & c_{33} \end{bmatrix} \quad .$$

For the case at hand, a material with a symmetry plane x_1x_2, the elastic constants c_{35} and c_{34} are zero as (1.37) shows. Thus,

$$\Gamma_{i\ell}(x_3) = \begin{bmatrix} c_{55} & c_{45} & 0 \\ c_{45} & c_{44} & 0 \\ 0 & 0 & c_{33} \end{bmatrix} \quad . \tag{3.10}$$

and (3.8) takes the form

$$\begin{vmatrix} c_{55} - \alpha & c_{45} & 0 \\ c_{45} & c_{44} - \alpha & 0 \\ 0 & 0 & c_{33} - \alpha \end{vmatrix} = 0$$

with $\alpha = \rho V^2$. This provides a cubic equation

$$(c_{33} - \alpha)\Big[(c_{55} - \alpha)(c_{44} - \alpha) - c_{45}^2\Big] = 0 \quad .$$

One root is obviously given by $\alpha_3 = c_{33}$ and $V_3 = \sqrt{c_{33}/\rho}$, whereas the other two phase velocities are associated with the roots of a quadratic equation,

$$\alpha_\kappa = \frac{c_{55}+c_{44}}{2} - \frac{(-1)^\kappa}{2}\left[\left(c_{44}-c_{55}\right)^2 + 4c_{45}^2\right]^{1/2} \tag{3.11}$$

$$\kappa = 1,2, \quad \alpha_\kappa = \rho V_\kappa^2 \quad .$$

To find the appropriate polarization we first substitute (3.10) and $V = V_3$ into (3.9) to obtain

$$(c_{55} - \rho V_3^2)A_1 + c_{45}A_2 = 0$$

$$c_{45}A_1 + (c_{44} - \rho V_3^2)A_2 = 0$$

$$(c_{33} - \rho V_3^2)A_3 = 0 \quad .$$

The first two of these relations are satisfied, provided $A_1 = A_2 = 0$, whereas the third one holds for any finite value of A_3. Thus, the displacement is polarized strictly along the x_3-axis, which is the propagation direction. In other words, the wave with the velocity $V_3 = \sqrt{c_{33}/\rho}$ is purely longitudinal.

Next, for the two other roots we get

$$(c_{55} - \rho V_\kappa^2)A_1 + c_{45}A_2 = 0$$

$$c_{45}A_1 + (c_{44} - \rho V_\kappa^2)A_2 = 0$$

$$(c_{33} - \rho V_\kappa^2)A_3 = 0 \quad .$$

Since $\rho V_\kappa^2 \neq c_{33}$ for $\kappa = 1,2$ the third equation provides $A_3 = 0$, indicating that the polarization is in the plane normal to x_3 and, thus, the associated waves are of pure shear modes. The first two equations then imply

$$\frac{A_2^{(\kappa)}}{A_1^{(\kappa)}} = \frac{(\rho V_\kappa^2 - c_{55})}{c_{45}} = \frac{c_{45}}{(\rho V_\kappa^2 - c_{44})} \quad .$$

This ratio also provides the angles β_1 and β_2, which the resulting displacement makes with the x_1-axis

$$\tan\beta_\kappa = A_2^{(\kappa)}/A_1^{(\kappa)} \quad ,$$

as may be seen from (3.2). The last two expressions along with (3.11) yield

$$\tan\beta_\kappa = d - (-1)^\kappa(d^2+1)^{1/2} \quad , \tag{3.12}$$

with $d = (c_{44} - c_{55})/(2c_{45})$. Figure 3-2 shows schematically the displacement vectors and associated velocities.

One observes that the symmetry of the structure imposes constraints on the possible motions. The intimate relation between structural properties and types of motion taking place is further illustrated by an example of cubic systems.

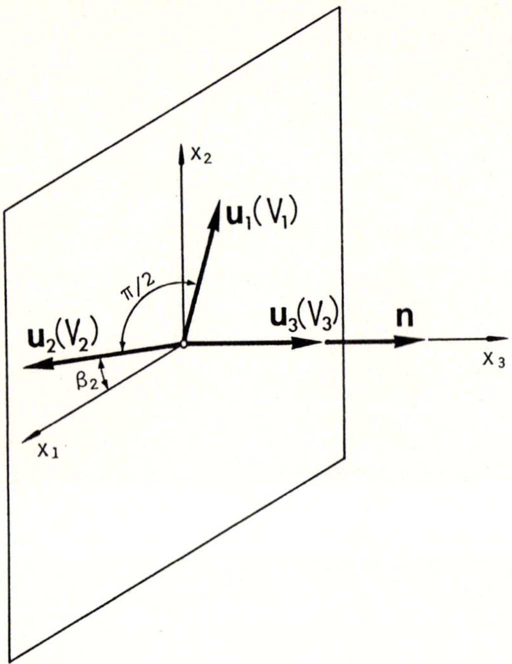

Fig. 3-2 Polarizations and velocities of waves in a medium with the x_1x_2-symmetry plane.

3.3 Pure Modes in Cubic Systems

The elastic constants, $c_{\alpha\beta}$, for a case of cubic symmetry are given in Figure 1-14 (see also (1.40)) provided the reference frame coincides with the symmetry axes, which is supposed to be the case here and henceforth. Substituting these into (3.7) one gets

$$\begin{aligned}\Gamma_{11} &= c_{11}n_1^2 + c_{44}(n_2^2 + n_3^2)\\ \Gamma_{12} &= \Gamma_{21} = (c_{12} + c_{44})n_1n_2\\ \Gamma_{13} &= \Gamma_{31} = (c_{12} + c_{44})n_1n_3\\ \Gamma_{22} &= c_{44}(n_1^2 + n_3^2) + c_{11}n_2^2\\ \Gamma_{32} &= \Gamma_{23} = (c_{12} + c_{44})n_2n_3\\ \Gamma_{33} &= c_{44}(n_1^2 + n_2^2) + c_{11}n_3^2 \quad .\end{aligned} \tag{3.13}$$

Since $n^2 = 1$, (3.9) can be rewritten as follows:

$$\begin{aligned}(\rho V^2 - c_{44})A_1 = (c_{11} - c_{44})n_1^2A_1 &+ (c_{12} + c_{44})n_1n_2A_2\\ &+ (c_{12} + c_{44})n_1n_3A_3\end{aligned}$$

$$
\begin{aligned}
(\rho V^2 - c_{44})A_2 &= (c_{12} + c_{44})n_1 n_2 A_1 + (c_{11} - c_{44})n_2^2 A_2 \\
&\quad + (c_{12} + c_{44})n_2 n_3 A_3 \\
(\rho V^2 - c_{44})A_3 &= (c_{12} + c_{44})n_1 n_3 A_1 + (c_{12} + c_{44})n_2 n_3 A_2 \\
&\quad + (c_{11} - c_{44})n_3^2 A_3 \quad .
\end{aligned}
\tag{3.14}
$$

To find a purely longitudinal mode one invokes the fact that for this wave the displacement $\boldsymbol{u}$ and the propagation vector $\boldsymbol{n}$ are related by $\boldsymbol{u} \times \boldsymbol{n} = 0$, or in coordinate form,

$$
\begin{aligned}
A_1 n_2 - A_2 n_1 &= 0 \\
A_2 n_3 - A_3 n_2 &= 0 \\
A_3 n_1 - A_1 n_3 &= 0 \quad .
\end{aligned}
$$

The above set of expressions is equivalent to the relation,

$$
n_1 : n_2 : n_3 = A_1 : A_2 : A_3 \quad . \tag{3.15}
$$

On the other hand, (3.14), governing wave propagation in a cubic system, provides

$$
\frac{n_1}{A + Bn_1^2} : \frac{n_2}{A + Bn_2^2} : \frac{n_3}{A + Bn_3^2} = A_1 : A_2 : A_3 \quad ,
$$

where $A = (\rho V^2 - c_{44})/c_{44}$ and $B = (c_{12} - c_{11} - 2c_{44})/c_{44}$. Comparison of the above expression and (3.15) shows that the directions $\boldsymbol{n}$, associated with purely longitudinal modes, must satisfy the relation,

$$
n_1 : n_2 : n_3 = \frac{n_1}{A + Bn_1^2} : \frac{n_2}{A + Bn_2^2} : \frac{n_3}{A + Bn_3^2} \quad . \tag{3.16}
$$

The above equation becomes an identity for

$$
B = 0 \rightarrow c_{44} = (c_{11} - c_{12})/2, \quad A \neq 0 \tag{3.17}
$$

indicating that in this case purely longitudinal waves run in any direction. Clearly, the entity b, defined as

$$
b = 2c_{44}/(c_{11} - c_{12})
$$

can be called the anisotropy factor. For isotropic media $b = 1$, as follows from (1.41), while deviations from this value describe the "strength" of anisotropy.

Next, consider the anisotropic case, $B \neq 0$, $(b \neq 1)$. It can readily be seen that (3.16) is satisfied in any one of the following three cases:

$$
\begin{aligned}
&\text{i)} \quad n_i = 1, \quad n_\kappa = n_\ell = 0 \\
&\text{ii)} \quad n_i = 0, \quad n_\kappa = n_\ell \quad i \neq \kappa \neq \ell \\
&\text{iii)} \quad n_i = n_\kappa = n_\ell \quad .
\end{aligned}
\tag{3.18}
$$

The directions given by these equations are shown in Figure 3-3. The directions [100], [010], and [001] satisfy the first relation, [110], [101], and [011] – the second, and [111] – the third. All these directions coincide with symmetry axes of a cubic system.

Similarly, the same directions allow for purely shear modes as well. In fact, the relation $\boldsymbol{u} \cdot \boldsymbol{n} = 0$, valid for S-waves, can be written as follows:

$$A_1 n_1 + A_2 n_2 + A_3 n_3 = 0 \quad . \tag{3.19}$$

Setting, for example, $n_2 = n_3 = 0$, $n_1 = 1$, one finds from (3.19) that $A_1 = 0$, whereas A_2 and A_3 remain arbitrary. It means that the polarization plane is the x_2x_3-plane, normal to [100]. Hence, the first of (3.18) yields the propagation direction of purely shear waves too. The remaining expressions can be examined in a similar way.

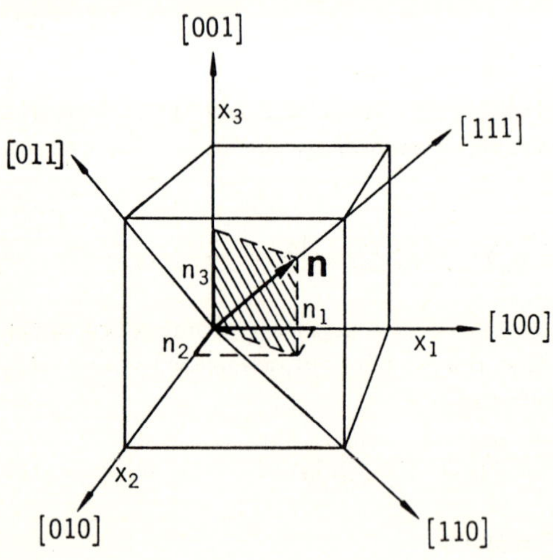

Fig. 3-3 Cubic System.

3.3.1 Propagation Along an Edge

Consider the procedure in more detail for the first of (3.18), setting, say,

$$n_1 = 1, \qquad n_2 = n_3 = 0 \quad .$$

Then, by virtue of (3.13)

$$\Gamma_{11} = c_{11}, \qquad \Gamma_{22} = c_{44}, \qquad \Gamma_{33} = c_{44} \tag{3.20}$$

$$\Gamma_{12} = \Gamma_{32} = \Gamma_{13} = 0$$

and, according to (3.8),

$$\begin{vmatrix} c_{11}-\rho V^2 & 0 & 0 \\ 0 & c_{44}-\rho V^2 & 0 \\ 0 & 0 & c_{44}-\rho V^2 \end{vmatrix} = 0 \quad .$$

The above equation has three roots

$$V_1 = \sqrt{c_{11}/\rho}, \qquad \text{and} \qquad V_2 = V_3 = \sqrt{c_{44}/\rho} \quad . \tag{3.21}$$

Equations (3.14) take the form

$$\begin{aligned} &(\rho V_\kappa^2 - c_{44})A_1 = (c_{11}-c_{44})A_1 \\ &(\rho V_\kappa^2 - c_{44})A_2 = 0 \qquad\qquad \kappa = 1,2,3 \\ &(\rho V_\kappa^2 - c_{44})A_3 = 0 \quad . \end{aligned} \tag{3.22}$$

This system admits two types of solution. The first is

$$A_1 \neq 0, \; A_2 = A_3 = 0, \qquad \text{and} \qquad V_1 = \sqrt{c_{11}/\rho} \quad , \tag{3.23}$$

which clearly represents a purely longitudinal mode, since the polarization is parallel to the prescribed direction, $n_1 = 1$, $n_2 = n_3 = 0$. The second type of solution is $A_1 = 0$, A_2 and A_3 are arbitrary and

$$V_2 = V_3 = \sqrt{c_{44}/\rho} \quad , \tag{3.24}$$

which represents purely shear waves. The analysis of wave propagation in the x_1-direction is, thus, completed.

3.3.2 Propagation Along a Face Diagonal

Now we shall consider the direction given by the second of (3.18), setting

$$n_1 = n_2, \quad n_3 = 0 \quad .$$

Since $n_1^2 + n_2^2 + n_3^2 = 1$, this means that

$$n_1 = n_2 = 1/\sqrt{2}, \quad n_3 = 0 \quad . \tag{3.25}$$

The propagation tensor, $\Gamma_{i\ell}$, is given, according to (3.13), by

$$\begin{aligned} &\Gamma_{11} = 1/2(c_{11}+c_{44}), && \Gamma_{12} = \Gamma_{21} = 1/2(c_{12}+c_{44}) \\ &\Gamma_{13} = \Gamma_{31} = 0, && \Gamma_{22} = 1/2(c_{11}+c_{44}) \\ &\Gamma_{32} = \Gamma_{23} = 0, && \Gamma_{33} = c_{44} \quad , \end{aligned}$$

whereas (3.14) become

$$\begin{aligned} &(\rho V^2 - 1/2c_{44} - 1/2c_{11})A_1 - 1/2(c_{12}+c_{44})A_2 = 0 \\ &-1/2(c_{12}+c_{44})A_1 + (\rho V^2 - 1/2c_{44} - 1/2c_{11})A_2 = 0 \\ &(\rho V^2 - c_{44})A_3 = 0 \quad . \end{aligned} \tag{3.26}$$

The determinantal equation is exactly of the type treated in Section 3.2. One may, therefore, apply the results obtained there, including (3.11). Adjusting notations, one gets the following velocities:

$$
\begin{aligned}
V_1 &= \sqrt{c_{44}/\rho} \\
V_2 &= [1/2(c_{11} - c_{12})/\rho]^{1/2} \\
V_3 &= [1/2(c_{11} + c_{12} + 2c_{44})/\rho]^{1/2} \quad .
\end{aligned}
\tag{3.27}
$$

The first root, $V = V_1$, relates to a shear wave, since setting $A_1 = A_2 = 0$ one satisfies (3.26), while A_3 remains an arbitrary value. Clearly, the displacement is normal to the propagation direction prescribed by (3.25). Substituting $A_3 = 0$ and $V = V_2$ into (3.26) yields $A_1 = -A_2$, which indicates that the associated mode is also shear. Finally, inserting $A_3 = 0$ and $V = V_3$, one gets $A_1 = A_2$, which defines a purely longitudinal wave. Figure 3-4 provides a graphical illustration of the situation.

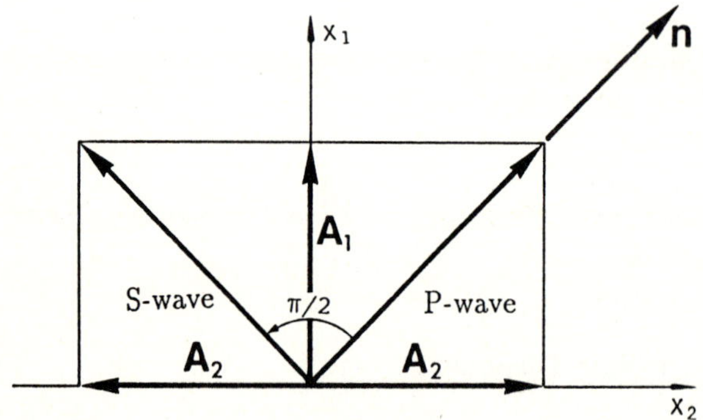

Fig. 3-4 Pure modes in a cubic system.

The same technique applies, of course, to the last of (3.18) as well. This section has dealt with the propagation of pure modes along the directions of symmetry axes of a cubic system. Now we turn to propagation in an arbitrary direction.

3.4 General Modes in Cubic Systems

Figure 3-5 shows that an arbitrary propagation direction, $\boldsymbol{n}$, is defined by two polar angles, θ and ϕ, as follows:

$$
n_1 = \sin\theta\cos\phi, \quad n_2 = \sin\theta\sin\phi, \quad n_3 = \cos\theta \quad . \tag{3.28}
$$

Having prescribed these angles, one may solve (3.14) and (3.8) to find the phase velocities and polarization. In general, a computer is needed to arrive at the final results. However, some of the cases admit an analytical treatment.

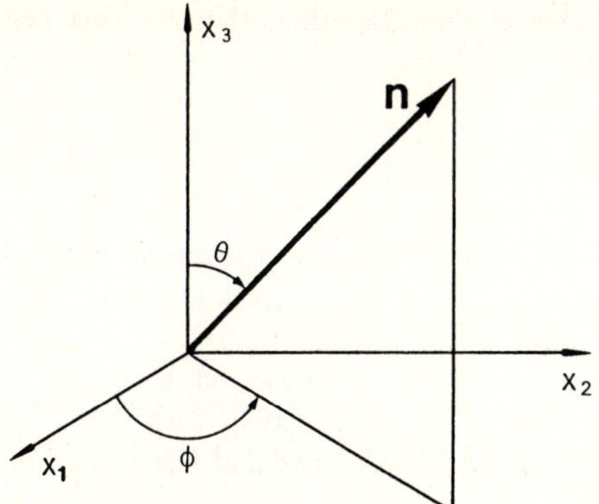

Fig. 3-5 Propagation direction, n, and the associated angles, θ and ϕ.

Let us consider, for example, the propagation in the x_1x_2-plane. Then $\theta = \pi/2$ and

$$n_1 = \cos\phi, \quad n_2 = \sin\phi, \quad n_3 = 0 \quad . \tag{3.29}$$

The determinantal equation (3.8) takes the form

$$|\Gamma_{i\ell} - \rho V^2\delta_{i\ell}| = \begin{vmatrix} \Gamma_{11} - \rho V^2 & \Gamma_{12} & 0 \\ \Gamma_{12} & \Gamma_{22} - \rho V^2 & 0 \\ 0 & 0 & \Gamma_{33} - \rho V^2 \end{vmatrix} = 0 \quad , \tag{3.30}$$

in which, according to (3.13),

$$\begin{aligned} \Gamma_{11} &= c_{11}\cos^2\phi + c_{44}\sin^2\phi, \quad \Gamma_{12} = (c_{12} + c_{44})\cos\phi\sin\phi \\ \Gamma_{22} &= c_{11}\sin^2\phi + c_{44}\cos^2\phi, \quad \Gamma_{33} = c_{44} \quad . \end{aligned} \tag{3.31}$$

The determinant is of the type investigated in Section 3.2. Thus, adjusting notations, one gets

$$\begin{aligned} V_3 &= \sqrt{\Gamma_{33}/\rho} = \sqrt{c_{44}/\rho} \\ \rho V_{1,2}^2 &= (\Gamma_{11} + \Gamma_{22})/2 \pm \left[(\Gamma_{11} - \Gamma_{22})^2 + 4\Gamma_{12}^2\right]^{1/2}/2 \quad . \end{aligned} \tag{3.32}$$

Making use of (3.31), the above expression is found to be

$$\rho V_{1,2}^2 = 1/2\Big\{c_{11} + c_{44} \pm \Big[(c_{11} - c_{44})^2\cos^2 2\phi + (c_{12} + c_{44})^2\sin^2 2\phi\Big]^{1/2}\Big\} \quad . \tag{3.33}$$

The velocity V_3, corresponds to a purely shear wave polarized in the x_3-direction, which can be verified with the help of Christoffel's equations, while V_1 and V_2, which are dependent on the angle ϕ, correspond to quasi-longitudinal and quasi-shear modes. Only for specific values of ϕ, namely $\phi = 0,\ \pi/2,\ \pi, ...,$

these modes convert to pure ones. For these specific directions the velocities are given by (3.33) as

$$\begin{aligned} V_1 &= \sqrt{c_{11}/\rho} \quad \text{(longitudinal wave)} \\ V_2 &= \sqrt{c_{44}/\rho} \quad \text{(shear wave)} \quad . \end{aligned} \tag{3.34}$$

It can be seen that the indicated directions coincide with the symmetry axes of a cubic system. In fact, the results given by (3.32) and (3.34) are identical with those of the previous section, given by (3.21).

To appreciate the effect of anisotropy on elastic response, let us compare the shear wave velocity for propagation along an edge, say, [100], with that of a diagonal direction, [110]. Making use of (3.33) with $\phi = \pi/4$ and the second of (3.34) one obtains

$$V_2^{[110]}/V_2^{[100]} = b^{-1/2} \quad ,$$

where b is the anisotropy factor defined in Section 3.3. For silicon, $b^{-1/2} \simeq 0.8$. For some of the cubic crystals $b \simeq 3$, which would bring the above ratio to about 0.6. Thus, ignoring the anisotropy effect can lead to considerable errors. Some typical values of elastic constants are given in Table 3.1.

Table 3.1 Elastic constants (cubic system)

Material	$\rho \cdot 10^{-3}$ (kg/m^3)	$c_{11} \cdot 10^{-10}$ (N/m^2)	$c_{12} \cdot 10^{-10}$ (N/m^2)	$c_{44} \cdot 10^{-10}$ (N/m^2)
Silicon	2.33	16.57	6.39	7.96
Copper	8.94	16.84	12.14	7.54
α-Iron	7.86	24.2	14.65	11.2
Bismuth-germanium oxide	9.23	12.8	3.05	2.55

This example indicates the effect of propagation direction on the elastic response. It can be conveniently displayed by the so-called slowness curves, constructed in polar coordinates, $r = 1/V$ and ϕ. These curves will be explained in more detail in Section 3.12. Figure 3-6 shows these curves, computed from (3.32) and (3.33), for silicon in the plane of the cube face. The values of the elastic constants and the density are those appearing in Table 3.1.

An analysis similar to that of Section 3.3 shows that the displacement vectors of quasi-longitudinal and quasi-shear waves, associated with the velocities given by (3.33), make angles β_1 and β_2 with the x_1-axis given by

$$\tan\beta_\kappa = d - (-1)^\kappa \sqrt{d^2+1}, \quad \kappa = 1,2 \quad ,$$

where $d = (\Gamma_{22} - \Gamma_{11})/(2\Gamma_{12})$. Given the elastic constants one can compute these angles as functions of ϕ via (3.31). Figure 3-7 shows the result for a quasi-longitudinal wave in silicon in the coordinates $\beta_1 - \phi$ and ϕ. The upper corner of the picture explains the notation. Thus, the direction of propagation, given by ϕ, markedly affects wave polarization.

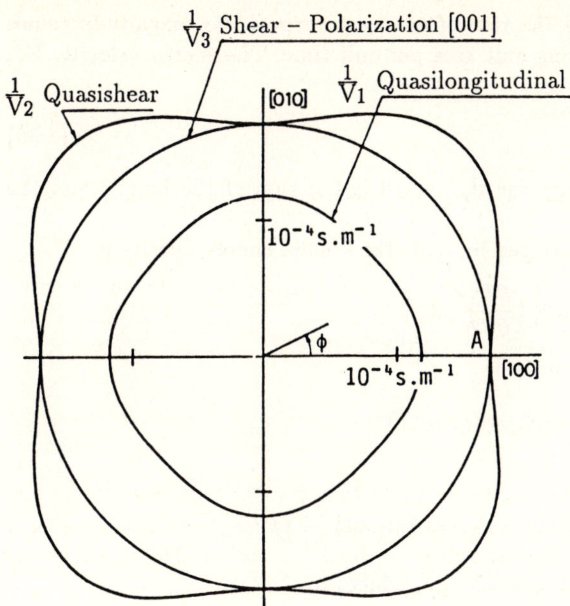

Fig. 3-6 Slowness curve for silicon.

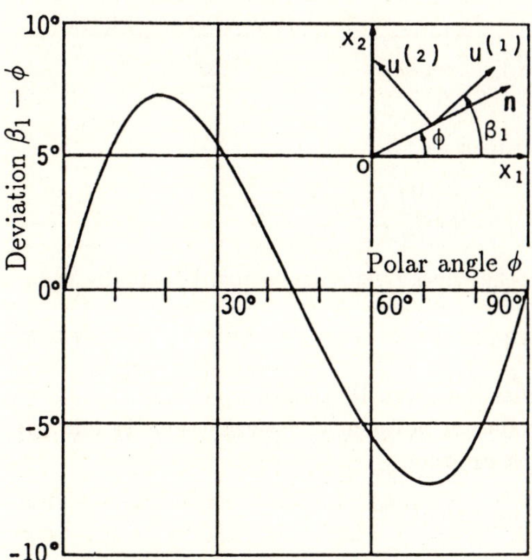

Fig. 3-7 Polarization vs. ϕ.

3.5 Energy Velocity

The Poynting vector, $\boldsymbol{P}$, introduced in Section 1.14 and given by

$$P_k = -\sigma_{ij}\dot{u}_j = -c_{ijk\ell}u_{\ell,k}\dot{u}_j \tag{3.35}$$

indicates the direction and the rate of energy transport. Its magnitude equals the amount of energy crossing unit area per unit time. The energy velocity, $\boldsymbol{V}^e$, is naturally defined as

$$V_k^e = P_k/E_o \quad , \tag{3.36}$$

where E_o is the total energy density, which is the sum of the kinetic and the potential energy densities.

For the plane waves described by (3.2) the kinetic energy density is

$$K = \rho(u_{i,t})^2/2 = \rho A_i^2\left(\frac{\partial f}{\partial \tau}\right)^2/2$$

and the potential energy density is

$$W = c_{ijk\ell}\varepsilon_{ij}\varepsilon_{k\ell}/2 = c_{ijk\ell}u_{i,j}u_{\ell,k}/2 \quad .$$

Since

$$u_{i,j} = -n_j A_i\left(\frac{\partial f}{\partial \tau}\right)/V, \quad u_{\ell,k} = -n_k A_\ell\left(\frac{\partial f}{\partial \tau}\right)/V \quad , \tag{3.37}$$

and in view of (3.3) the potential energy reduces to

$$W = \rho A_i^2\left(\frac{\partial f}{\partial \tau}\right)^2/2 \quad .$$

Thus, the total energy density is

$$E_o = K + W = \rho A_i^2\left(\frac{\partial f}{\partial \tau}\right)^2 \quad ,$$

while the Poynting vector can be written as

$$P_i = -c_{ijk\ell}u_{\ell,k}\dot{u}_j = c_{ijk\ell}A_jA_\ell n_k\left(\frac{\partial f}{\partial \tau}\right)^2/V \quad ,$$

where (3.37) has been used. Hence, the energy velocity vector, V_i^e, finally is

$$V_i^e = c_{ijk\ell}A_jA_\ell n_k/(\rho A_i^2 V) \quad , \tag{3.38}$$

where use has been made of the definition given by (3.36).

The question arises about the relation between the energy velocity and the phase velocity. To this end, write the product

$$\boldsymbol{V}^e \cdot \boldsymbol{n} = V_i^e n_i = c_{ijk\ell}A_jA_\ell n_k n_i/(\rho V A_i^2) \quad ,$$

where $\boldsymbol{n}$ is the propagation direction vector. In view of Christoffel's equation (3.3) this results in

$$\boldsymbol{V}^e \cdot \boldsymbol{n} = V \quad . \tag{3.39}$$

Thus, the component of the energy velocity vector in the propagation direction equals the phase velocity of the plane wave. Many authors define the propagation mode as a "pure", if the vectors $\boldsymbol{V}^e$ and $\boldsymbol{n}$ are parallel, unlike the definition given in Section 3.1.

Next, compare the energy velocity with the group velocity introduced in Section 1.16. Extending (1.99) to the three dimensional case, one defines the vector of group velocity, $\boldsymbol{V}^g$, as

$$V_j^g = \frac{\partial \omega}{\partial k_j} \quad , \tag{3.40}$$

where $k_j = kn_j$.

To actually compute this value one needs the function $\omega = \tilde{f}(k_j)$. It can be shown that V and n_j are related by the same function, $V = \tilde{f}(n_j)$. Indeed, multiplication of (3.6) by k^6 yields

$$\begin{aligned} & k^6 |c_{ijk\ell} n_j n_k - \rho V^2 \delta_{i\ell}| \\ & = |c_{ijk\ell} k n_j k n_k - \rho k^2 V^2 \delta_{i\ell}| \\ & = |c_{ijk\ell} k_j k_k - \rho \omega^2 \delta_{i\ell}| = 0 \quad . \end{aligned}$$

Comparison of the first and last expressions shows that these functions are governed by the same equation. Hence,

$$V_j^g = \frac{\partial \omega}{\partial k_j} = \frac{\partial V}{\partial n_j} \quad .$$

On the other hand,

$$\frac{\partial V^2}{\partial n_j} = 2V \frac{\partial V}{\partial n_j} = 2 c_{ijk\ell} \frac{n_k A_i A_\ell}{\rho A_i^2} \quad ,$$

as follows from (3.3). Hence,

$$V_j^g = \frac{\partial V}{\partial n_j} = c_{ijk\ell} n_k A_i A_\ell / (\rho A_i^2 V) \quad , \tag{3.41}$$

which coincides with the right-hand side of (3.38). Thus, the energy velocity is equal to the group velocity for the considered case of plane elastic waves. This enables one to simplify the analysis by computing $\boldsymbol{V}^g$ in place of $\boldsymbol{V}^e$, which is more convenient, as has been shown in the foregoing.

The concept of the energy velocity has evident practical applications. Consider, for example, a wave packet radiated by a plane wave transducer (Figure 3-8). While the wave front travel in the direction $\boldsymbol{n}$, normal to the transducer surface, the energy propagates in the direction $\boldsymbol{V}^e = \boldsymbol{V}^g$, in other words, along the acoustic ray, which may be at an angle to $\boldsymbol{n}$. Hence, the receiving transducer should be shifted in order to intercept efficiently the wave packet.

3.6 Waves in Piezoelectric Media

We go over to piezoelectric media, unique features of which have extensive engineering applications. Clearly, Maxwell's equations must be incorporated in addition to the constitutive law given in Section 1.9 and the equations of motion.

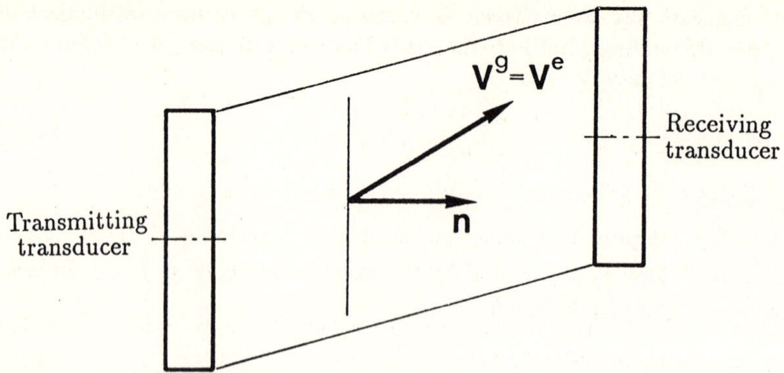

Fig. 3-8 Shift of the receiving transducer according to the energy velocity.

However, the exact analysis of piezoelectric waves may be extremely tedious. Fortunately, the basic equations admit a radical simplification based on a possibility of neglecting the time-dependent magnetic field caused by elastic stresses via the piezoelectricity mechanism. This will be considered in more detail in the next section, while in what follows here we deal with a rigorous treatment of a piezoelectric response, making use of a simple example of a cubic system.

For this case of symmetry the elastic constants, $c_{\alpha\beta}$, the dielectric constants, ζ_{ij}, and the piezoelectric constants, $f_{i\alpha}$, are displayed in Figure 1-14. It follows that the last two tensors reduce to ζ_{11}, and f_{14}, respectively.

Consider a plane shear wave

$$u_1 = Ae^{i(\omega t - kx_2)} \quad , \tag{3.42}$$

where ω is a stipulated value and the wave number, k, should be found from the governing equations. This would provide the phase velocity, $V = \omega/k$. Equation (3.42) implies the following shear strain

$$\varepsilon_{12} = \varepsilon_6 = u_{1,2} = -Aike^{i(\omega t - kx_2)} \quad . \tag{3.43}$$

For a non-magnetic medium with no conductive losses and no current sources Maxwell's equations are

$$-\nabla \times \boldsymbol{E} = \mu_o \dot{\boldsymbol{H}}, \quad \nabla \times \boldsymbol{H} = \dot{\boldsymbol{D}} \quad , \tag{3.44}$$

where $\boldsymbol{E}$ and $\boldsymbol{H}$ are the electric and magnetic fields (intensities), respectively, μ_o is the permeability. Also, the electric displacement, $\boldsymbol{D}$, and $\boldsymbol{E}$ are related by*

$$\boldsymbol{D} = \bar{\zeta}\boldsymbol{E} \quad , \tag{3.45}$$

where $\bar{\zeta}$ is the tensor of dielectric constants.

* For standard notations see the IEEE Standard on Piezoelectricity and Mariani (1985).

It is natural to assume that the electromagnetic waves run also in the x_2-direction and $\partial/\partial x_1 = \partial/\partial x_3 = 0$ in (3.44). Then we get that

$$E_{3,2} = -\mu_o \dot{H}_1 \quad (a) \qquad H_{3,2} = \dot{D}_1 \quad (c)$$
$$E_{1,2} = \mu_o \dot{H}_3 \quad (b) \qquad H_{1,2} = -\dot{D}_3 \quad (d) \tag{3.46}$$

and that H_2 and D_2 are time-independent and may be set to zero.

On invoking (1.58) we obtain

$$D_3 = \zeta_{11}^{\varepsilon} E_3 + f_{14}\varepsilon_6 \quad ,$$

where the second term represents the contribution of the elastic field to the electric one. Equations (3.46a,d) and the above relation provide

$$H_{1,2} = -\zeta_{11}^{\varepsilon}\dot{E}_3 - f_{14}\dot{\varepsilon}_6$$

and

$$E_{3,22} - \mu_o \zeta_{11}^{\varepsilon} E_{3,tt} = \mu_o f_{14} \varepsilon_{6,tt} \quad , \tag{3.47}$$

which is a nonhomogeneous wave equation with the forcing term depending on the piezoelectric constant, f_{14}.

The steady-state solution of (3.47) and (3.43) is given by

$$E_3 = -\mu_o f_{14}\omega^2 \varepsilon_6 / (\mu_o \zeta_{11}^{\varepsilon}\omega^2 - k^2) \quad . \tag{3.48}$$

Now (1.59) yields the associated stress

$$\sigma_{12} = \sigma_6 = \left[c_{44}^E + \mu_o f_{14}^2 \omega^2 / (\mu_o \zeta_{11}^{\varepsilon} \omega^2 - k^2) \right] \varepsilon_6 \quad , \tag{3.49}$$

where use has been made of (1.40) to specify the elastic contribution. The superscripts E and ε denote constant electric field and constant strain, respectively.

The equation of motion (1.15) with the body forces neglected

$$\sigma_{12,2} = \rho \ddot{u}_1$$

provides the dispersion relation

$$F_1(V) = (\rho\omega^2 - c_{44}^E k^2)(\mu_o \zeta_{11}^{\varepsilon}\omega^2 - k^2) = \mu_o f_{14}^2 \omega^2 k^2 = F_2(V) \quad , \tag{3.50}$$

which defines the phase velocities $V = \omega/k$.

Note that for $f_{14} = 0$ we get two uncoupled waves with the velocities

$$V_1^2 = c_{44}/\rho \qquad \text{and} \qquad V_2^2 = (\mu_o \zeta_{11}^{\varepsilon})^{-1} \quad , \tag{3.51}$$

as expected. First of these velocities is of a purely elastic wave and the second of an electromagnetic wave.

If $f_{14} \neq 0$, there exist coupled elastic and electromagnetic fields propagating with the speeds V_{pz1} and V_{pz2}, respectively, defined by the roots of (3.50). The locations of the roots are seen from Figure 3-9, where the notations $F_1(V)$ and $F_2(V)$ should be clear from (3.50). Hence,

$$V_{pz1} < V_1 \qquad \text{and} \qquad V_{pz2} > V_2 \quad .$$

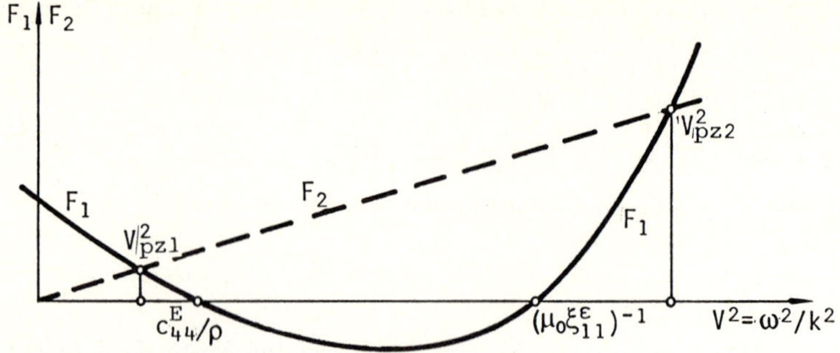

Fig. 3-9 Roots of (3.50).

Computations show that for typical values of the constants appearing in (3.50) the shifts, indicated by the above inequalities, are small.

3.7 Quasi-Electrostatic Approximation of Piezoelectricity

The piezoelectric waves may be separated into two constituents by taking into account that the ratio of velocities of electromagnetic disturbances to elastic ones is about 10^5 (for a solid medium). The wavelength of elastic waves is thus much smaller than that of electromagnetic waves of the same frequency and these two phenomena have distinctly different scales. It follows that the magnetic effects caused by the electric field which is generated by elastic strains can be neglected. This constitutes the quasi-electrostatic approximation, which assumes that the time-dependent magnetic field may be disregarded, but, at the same time, accounts for the dynamic nature of the elastic field.

Under the above assumption we set $\dot{\boldsymbol{H}} = 0$ and (3.44) yields

$$\nabla \times \boldsymbol{E} = 0 \quad .$$

This expression motivates a resort to the scalar potential, T, defined as follows:

$$\boldsymbol{E} = -\operatorname{\mathbf{grad}} T \quad . \tag{3.52}$$

Now invoke (1.57) to derive in the sequel the equations governing the piezoelectric media in the above approximation,

$$\sigma_{ij} = c^E_{ijk\ell}\epsilon_{k\ell} - f_{kij}E_k \quad , \tag{3.53}$$

which relates the stresses, σ_{ij}, the strains, $\epsilon_{k\ell}$, and the electric field, E_k. Since $c_{ijk\ell}\epsilon_{k\ell} = c_{ijk\ell}u_{\ell,k}$ (see (1.31)), and in view of (3.52) the stresses can be put in the form

$$\sigma_{ij} = c_{ijk\ell}u_{\ell,k} + f_{kij}T_{,k} \quad .$$

Inserting this representation into the equations of motion,

$$\rho\ddot{u}_i = \sigma_{ij,j}$$

one obtains

$$\rho \ddot{u}_i = c^E_{ijk\ell} u_{\ell,jk} + f_{kij} T_{,jk} \quad . \tag{3.54}$$

This equation, however, cannot be solved, since it involves two unknown functions. The equation of direct piezoelectric effect (1.56) given by

$$D_j = \zeta^{\varepsilon}_{jk} E_k + f_{jk\ell} \varepsilon_{k\ell}$$

is therefore utilized to relate the electrical displacement, D_i, the electric field, E_j, and the strains, ε_{jk}. Here ζ^{ε}_{ij} stands for the dielectric constants under the constant strains. Making use of (3.52) and replacing the strains by the displacement gradients the electrical displacement can be rewritten as

$$D_j = -\zeta^{\varepsilon}_{jk} T_{,k} + f_{jk\ell} u_{\ell,k} \quad . \tag{3.55}$$

On invoking the equation for insulators,

$$D_{j,j} = 0$$

one gets from (3.55)

$$f_{jk\ell} u_{\ell,jk} - \zeta^{\varepsilon}_{jk} T_{,jk} = 0 \quad . \tag{3.56}$$

Two relations (3.54) and (3.56) constitute a closed-form system with respect to the functions $\boldsymbol{u}$ and T. Assume the solution to these equations in the form of running waves

$$u_i = A_i f(\tau) \qquad \text{and} \qquad T = BF(\tau) \quad , \tag{3.57}$$

where τ is the local time given by

$$\tau = t - \boldsymbol{n} \cdot \boldsymbol{r} / V = t - n_i x_i / V \quad .$$

The derivatives of interest are

$$\ddot{u}_i = A_i \partial^2 f / \partial \tau^2, \quad E_j = -T_{,j} = n_j B (\partial F / \partial \tau) / V$$

$$u_{\ell,jk} = A_\ell n_j n_k (\partial^2 f / \partial \tau^2) / V^2$$

$$T_{,jk} = B n_j n_k (\partial^2 F / \partial \tau^2) / V^2 \quad .$$

Now (3.54) and (3.56) yield

$$\rho V^2 A_i = \Gamma^E_{i\ell} A_\ell + \gamma_i B \tag{3.58}$$

$$\gamma_\ell A_\ell - \zeta_o B = 0 \quad ,$$

with the following abbreviations:

$$\Gamma_{i\ell} = c^E_{ijk\ell} n_j n_k, \quad \gamma_i = f_{kij} n_j n_k \tag{3.59}$$

and

$$\zeta_o = \zeta^{\varepsilon}_{j\kappa} n_j n_\kappa \quad . \tag{3.60}$$

As (3.59) and (3.60) indicate, $\Gamma_{i\ell}$ is the second rank tensor similar to that of non-piezoelectric media, γ_i is vector and ζ_o is scalar. It would be useful for the following considerations to write down the explicit expansion for γ_i

$$\gamma_i = f_{1i1}n_1^2 + f_{2i2}n_2^2 + f_{3i3}n_3^2 + (f_{1i2} + f_{2i1})n_1n_2 + (f_{1i3} + f_{3i1})n_1n_3 + (f_{2i3} + f_{3i2})n_2n_3 \quad . \tag{3.61}$$

In the convenient matrix notation, introduced in Section 1.7, the above expression is

$$\begin{aligned}\gamma_1 &= f_{11}n_1^2 + f_{26}n_2^2 + f_{35}n_3^2 + (f_{16} + f_{21})n_1n_2 \\ &\quad + (f_{15} + f_{31})n_1n_3 + (f_{25} + f_{36})n_2n_3 \\ \gamma_2 &= f_{16}n_1^2 + f_{22}n_2^2 + f_{34}n_3^2 + (f_{12} + f_{26})n_1n_2 \\ &\quad + (f_{14} + f_{36})n_1n_3 + (f_{24} + f_{32})n_2n_3 \\ \gamma_3 &= f_{15}n_1^2 + f_{24}n_2^2 + f_{33}n_3^2 + (f_{14} + f_{25})n_1n_2 \\ &\quad + (f_{13} + f_{35})n_1n_3 + (f_{23} + f_{34})n_2n_3 \quad .\end{aligned} \tag{3.62}$$

Returning to (3.58) and eliminating B we get

$$\rho V^2 A_i = (\Gamma_{i\ell} + \gamma_i\gamma_\ell/\zeta_o)A_\ell \quad .$$

Since $A_i = \delta_{i\ell}A_\ell$ this expression can be put in the form of Christoffel's equation

$$(\tilde{\Gamma}_{i\ell} - \rho V^2\delta_{i\ell})A_\ell = 0 \quad , \tag{3.63}$$

where $\tilde{\Gamma}_{i\ell}$ is the propagation tensor for piezoelectric media defined by

$$\tilde{\Gamma}_{i\ell} = \Gamma_{i\ell} + \gamma_i\gamma_\ell/\zeta_o \quad . \tag{3.64}$$

As expected, $\tilde{\Gamma}_{i\ell}$ reduces to $\Gamma_{i\ell}$ if we neglect the piezoelectric effect.

The analysis of plane waves in the framework of the quasistatic approximation of piezoelectricity is thus similar to that of non-piezoelectric media, provided the propagation direction and material constants are given. The determinantal equation associated with (3.63) is

$$|\tilde{\Gamma}_{i\ell} - \rho V^2\delta_{i\ell}| = 0 \quad ,$$

which again enables us to find the wave speeds.

The results obtained will be used in the next section to investigate a piezoelectric cubic system and to discover thus the specific effects of piezoelectricity on waves in anisotropic media.

3.8 Modes in a Piezoelectric Cubic System

As an example, we consider propagation in the x_1x_2-plane (see Figure 3-5). Then $\theta = \pi/2$, and (3.28) yields

$$n_1 = \cos\phi, \qquad n_2 = \sin\phi, \qquad \text{and} \qquad n_3 = 0 \quad . \tag{3.65}$$

The elastic, piezoelectric, and dielectric constants of a cubic system should be clear from Figure 1-14. Some typical values of piezoelectric and dielectric constants are given in Table 3.2.

Table 3.2 Piezoelectric and dielectric constants (cubic system)

Material	Class	Piezoelectric constant (C/m^2)	Dielectric constant $(10^{-11}F/m)$
Gallium arsenide	$\bar{4}3m$	$f_{14} = -0.16$	$\zeta_{11}^{\varepsilon} = 9.73$
Bismuth germanium oxide	23	$f_{14} = 0.99$	$\zeta_{11}^{\varepsilon} = 34.2$

Inserting the piezoelectric moduli into (3.62) we get

$$\gamma_1 = 2f_{14}n_2n_3, \quad \gamma_2 = 2f_{14}n_1n_3, \quad \gamma_3 = 2f_{14}n_1n_2 \quad .$$

In view of (3.65)

$$\gamma_1 = \gamma_2 = 0, \qquad \gamma_3 = f_{14}\sin 2\phi \quad . \tag{3.66}$$

Equation (3.64) shows that only one component of $\tilde{\Gamma}_{i\ell}$ differs from its non-piezoelectric analogue, namely, $\tilde{\Gamma}_{33}$, which is given by

$$\tilde{\Gamma}_{33} = \Gamma_{33} + \gamma_3^2/\zeta_o \quad , \tag{3.67}$$

whereas the other components are identical to those of the previously investigated case of non-piezoelectric cubic system, namely

$$\tilde{\Gamma}_{11} = \Gamma_{11}, \quad \tilde{\Gamma}_{22} = \Gamma_{22}, \quad \tilde{\Gamma}_{12} = \Gamma_{12},$$
$$\tilde{\Gamma}_{13} = \Gamma_{13}, \quad \tilde{\Gamma}_{23} = \Gamma_{23} \quad . \tag{3.68}$$

Moreover, (3.64) and (3.65) show that $\Gamma_{13} = \Gamma_{23} = 0$ and, hence, the tensor, $\tilde{\Gamma}_{i\ell}$, is given by

$$\tilde{\Gamma}_{i\ell} = \begin{bmatrix} \Gamma_{11} & \Gamma_{12} & 0 \\ \Gamma_{12} & \Gamma_{22} & 0 \\ 0 & 0 & \tilde{\Gamma}_{33} \end{bmatrix} \quad .$$

Accordingly, the analysis of Section 3.5 applies herein. There are three types of waves: quasi-longitudinal and quasi-shear with the velocities V_1 and V_2, respectively, and a purely shear wave with the velocity V_3. The values of V_1 and V_2 do not depend on Γ_{33}, as the second of (3.32) shows. Hence, these modes are not affected by the piezoelectric effect. On the other hand, for the shear mode polarized along [001], one gets, replacing Γ_{33} in the first of (3.32), by $\tilde{\Gamma}_{33}$

$$V_3 = (\tilde{\Gamma}_{33}/\rho)^{1/2} = \left[(\Gamma_{33} + \gamma_3^2/\zeta_o)/\rho\right]^{1/2} \quad .$$

Making use of (3.31), (3.66) and (3.60) this expression may be put as follows:

$$V_3 = (c_{44}^E/\rho)^{1/2}\left(1 + \frac{f_{14}^2}{\zeta_{11}^{\varepsilon}c_{44}^E}\sin^2 2\phi\right)^{1/2} \quad . \tag{3.69}$$

Thus, this time the velocity of the pure shear mode depends on the angle ϕ. This phenomena has not been observed in non-piezoelectric materials. Figure 3-10 shows the slowness curves for bismuth-germanium oxide ($Bi_{12}GeO_{20}$) in the (001) plane. Relevant constants are given in Tables 3.1 and 3.2. The purely shear mode is shown by the solid line for piezoelectric material and by the broken one for non-piezoelectric.

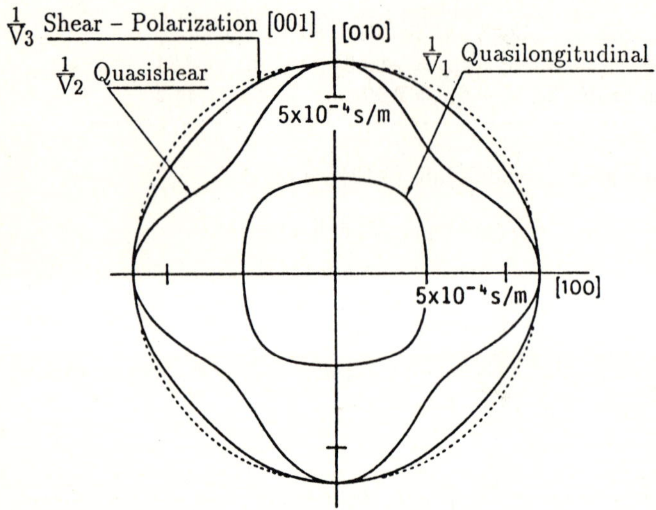

Fig. 3-10 Slowness curves for bismuth-germanium oxide.

3.9 Bulk Wave Piezoelectric Transducer

Figure 3-11 shows schematically a piezoelectric transducer used to launch bulk elastic waves. The electric field, applied to the electrodes, causes vibrations of the piezoelectric material. The associated disturbance crosses the inner electrode and enters the propagation medium in the form of a beam. In the near-field the beam is controlled by the size of the electrodes, but it spreads out while propagating inside the medium. Either longitudinal or shear modes can be launched by a proper choice of piezoelectric material and its orientation with respect to the propagation medium.

The exact and comprehensive analysis of the transducer is complicated and time consuming. It is possible to simplify the calculations provided the electrode thickness is small compared to the length of the excited wave. Under this condition the transducer can be viewed as consisting only of a piezoelectric layer and propagation material as shown in Figure 3-12.

The generation of longitudinal waves will be considered in what follows by applying the one-dimensional approximation. Since the propagation medium is

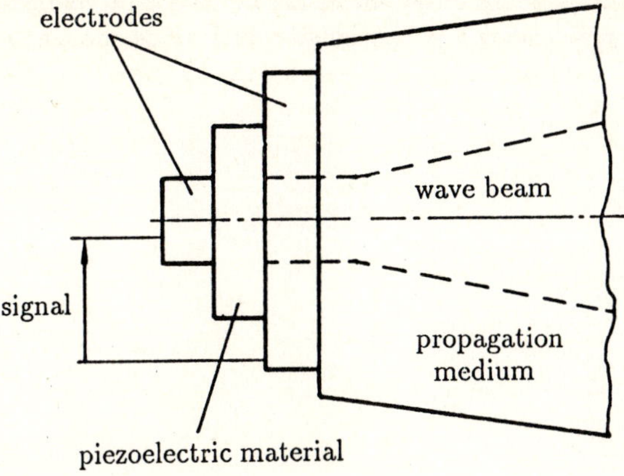

Fig. 3-11 Piezoelectric transducer.

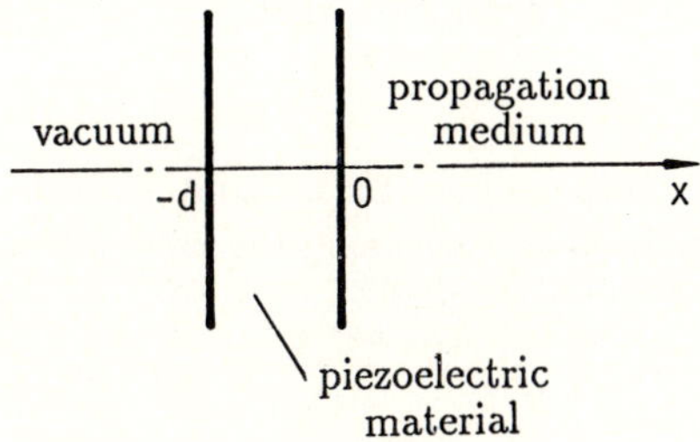

Fig. 3-12 Simplified version of Figure 3-11.

treated as infinite, no reflected waves are involved. Denote the displacement in the medium as $u(x,t)$, and that in the piezoelectric material as $u_p(x,t)$, where x designates the direction of propagation. The stresses are denoted as $\sigma(x,t)$ and $\sigma_p(x,t)$.

The piezoelectric material is bonded to the propagation medium at $x = 0$ and, on the other hand, no stress is present at the free plane $x = -d$. The following boundary conditions reflect these circumstances

$$\begin{aligned} u_p(0,t) &= u(0,t) \\ \sigma_p(0,t) &= \sigma(0,t) \\ \sigma_p(-d,t) &= 0 \quad . \end{aligned} \tag{3.70}$$

To determine the orientation needed to launch, say, a purely longitudinal wave running along the x-direction, let us write down the appropriate Christoffel equation (3.63), setting $i = 1$, $(x = x_1)$:

$$(\tilde{\Gamma}_{11} - \rho V^2)A_1 + \tilde{\Gamma}_{12}A_2 + \tilde{\Gamma}_{13}A_{13} = 0 \quad .$$

Each term represents the x_1-component of the forces acting on a unit volume of piezoelectric material. Clearly, to ensure a purely longitudinal mode one may set

$$\tilde{\Gamma}_{12} = \tilde{\Gamma}_{13} = 0 \quad .$$

Since

$$\tilde{\Gamma}_{i\ell} = \Gamma_{i\ell} + \gamma_i \gamma_\ell / \zeta_o \quad ,$$

this implies

$$\Gamma_{12} = \Gamma_{13} = 0 \tag{3.71}$$

and, consequently,

$$\gamma_2 = \gamma_3 = 0 \quad . \tag{3.72}$$

To satisfy (3.71) when $n_1 = 1$, $n_2 = n_3 = 0$, one sets

$$c_{16} = c_{15} = 0 \tag{3.73}$$

as (3.7) indicates. On the other hand, (3.72) and (3.62) show that

$$f_{16} = f_{15} = 0 \quad . \tag{3.74}$$

Equations (3.73) and (3.74) provide the information about the orientation of the piezoelectric crystal needed to launch a longitudinal wave along the x-direction.

3.10 Generation of Harmonic Waves. Resonators

The propagation of harmonic waves is of primary interest. The displacements and the stresses are therefore represented by

$$u(x,t) = u(x)e^{-i\omega t}, \qquad u_p(x,t) = u_p(x)e^{-i\omega t}$$
$$\sigma(x,t) = \sigma(x)e^{-i\omega t}, \qquad \sigma_p(x,t) = \sigma_p(x)e^{-i\omega t},$$

where $x = x_1$. This is obviously a one-dimensional approximation. Since the system is linear, the multiplier $e^{-i\omega t}$ will be frequently omitted in what follows.

Let us consider the electrical displacement, the electrical field and the stress in the piezoelectric material. The equation, $D_{i,i} = 0$, reduces to

$$\partial D / \partial x = 0 \quad ,$$

which implies

$$D = \text{const} \quad .$$

From (3.55) and (3.52) one gets

$$D = f \partial u_p(x) / \partial x + \zeta E \quad , \tag{3.75}$$

where f and ζ denote the piezoelectric and the dielectric constants. Introducing the potential difference $v = V_o e^{-i\omega t}$, which is applied to the electrodes, and integrating (3.75), we get

$$\int_0^{-d} D dx = -Dd = -\zeta V_o + f[u_p(-d) - u_p(0)] \quad , \tag{3.76}$$

where

$$V_o = -\int_0^{-d} E dx \quad .$$

On the other hand, the stress is given by (3.53) as

$$\sigma_p(x) = c_p \partial u_p / \partial x - fE \tag{3.77}$$

with c_p denoting the stiffness of the piezoelectric material at constant E. Eliminating E from (3.75) and (3.77) one obtains the relation between σ_p and D,

$$\sigma_p(x) = c_e \partial u_p / \partial x - fD/\zeta \quad , \tag{3.78}$$

where the effective stiffness c_e is given by

$$c_e = c_p + f^2/\zeta \quad . \tag{3.79}$$

Since the piezoelectric material is in a state of steady-state vibrations, its displacement represents a standing wave solution

$$u_p(x) = a_p e^{ik_p x} + b_p e^{-ik_p x} \quad , \tag{3.80}$$

where a_p, b_p are as yet unknown, whereas k_p is the wave number. The associated stress is given, according to (3.77), by

$$\sigma_p(x) = ik_p c_p (a_p e^{ik_p x} - b_p e^{-ik_p x}) - fE \quad . \tag{3.81}$$

The propagation medium is taken to be infinitely extended and a proper approximation is one of progressive waves

$$u = ae^{ikx} \tag{3.82}$$

$$\sigma = ik(\lambda + 2\mu)ae^{ikx} \quad , \tag{3.83}$$

where a is unknown amplitude and k is the wave number. To find the coefficients appearing in the expressions for the stress and the strain one employs the boundary conditions given by (3.70). Since the number of unknowns equals the number of equations, the coefficients can always be determined. In particular, the amplitude of the wave emitted into the propagation medium, a, is found to be

$$a = -2ifV_o/(d\omega m Z) \quad , \tag{3.84}$$

where

$$m = (\cos\alpha - iZ_p \sin\alpha/Z)/\sin^2(\alpha/2)$$

$$\alpha = k_p d$$

and $Z_p = k_p c_e/\omega$, $Z = k(\lambda + 2\mu)/\omega$ are the impedances of the two media involved.

The average power flux (see (2.133)) is now estimated by means of the relation

$$\langle I \rangle = Z\omega^2 |a^2| A/2 \quad ,$$

where A is the emitted beam cross-section. The above relation follows also immediately from (2.137) by adjusting the notation for the amplitude, a.

Upon substitution of (3.84) one gets

$$\langle I \rangle = 2A(\lambda + 2\mu) f^2 V_o^2 / (d^2 Z^2 |m|^2 c_\alpha) \quad . \tag{3.85}$$

In (3.85) only the term $|m|^2$ is frequency-dependent, which enables one to investigate the frequency effect on radiated power in a simple way.

In particular, it follows that $\langle I \rangle \to 0$ when $|m|^2 \to \infty$. The expression for m shows that this case takes place when

$$\sin^4(\alpha/2) = 0$$

$$\alpha = k_p d = (2\pi d)/\lambda_p = 2n\pi, \quad n = d/\lambda_p = 1, 2, 3 \ldots$$

Thus, it is necessary to avoid the thickness, d, of the piezoelectric plate which is equal to an integral number of wavelengths.

A more elaborate analysis shows that the dependence of the radiated power on frequency, given by (3.85), is strongly influenced by the ratio, Z_p/Z. When $Z_p/Z > 1$ the transducer has actually both its faces nearly free and operates at half-wavelength, $\lambda_p/2 = d$. Conversely, when $Z_p/Z < 1$, it has one of its faces almost firmly bonded, while the other is free, which changes operating conditions.

3.11 Improved Analysis of Resonators

The above method holds for low-frequency waves only (below 100 MHz) due to neglect of the electrode thickness. However, it extends to a more general case, shown in Figure 3-11 and then in a more detailed way in Figure 3-13.

Denote the values related to the earth electrode by subscript a, to the piezoelectric solid by p, and to the external electrode by e. Then the displacements may be approximated in a way similar to (3.80)

$$u_n = (a_n e^{ik_n x} + b_n e^{-ik_n x}) e^{-i\omega t} \quad , \tag{3.86}$$

where n takes the values a, p, and e. Accordingly, the stresses are given by

$$\sigma_n = \left[ik_n c_n (a_n e^{ik_n x} - b_n e^{-ik_n x}) - \delta_{np} f E\right] e^{-i\omega t} \quad , \tag{3.87}$$

where c_n denotes an appropriate elastic stiffness.

This expression indicates that the fE-term appears for the piezoelectric solid only. The displacement and the stress in the propagation medium are given by (3.82) and (3.83). There are seven unknown coefficients which can be found from

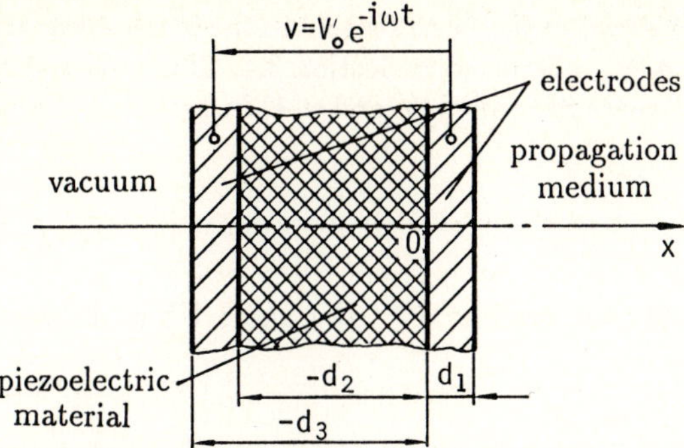

Fig. 3-13 Geometry of a transducer.

boundary conditions expressing the continuity of displacements and stresses (see Figure 3-13)

$$\begin{aligned} u_a(d_1) &= u(d_1), & \sigma_a(d_1) &= \sigma(d_1) \\ u_p(0) &= u_a(0), & \sigma_p(0) &= \sigma_a(0) \\ u_e(-d_2) &= u_p(-d_2), & \sigma_e(-d_2) &= \sigma_p(-d_2) \end{aligned} \tag{3.88}$$

and the condition at the free surface of the external electrode, which is given by

$$\sigma_e(-d_3) = 0 \quad . \tag{3.89}$$

The seven equations (3.88) and (3.89) enable one to determine the unknown coefficients, the expressions of which are, however, too lengthy to be written here. The radiated power shows, this time, a much more complicated dependence on frequency and structural parameters than in the previous case. A relevant point to note is that the presented one-dimensional theory is essentially an approximation, which, in particular, neglects deformations due to Poisson's effect. A model of a piezoelectric plate could yield far better results at the expense of a more involved analysis.

3.12 Characteristic Surfaces

To complete this chapter we consider some geometric surfaces, which possess quite remarkable features and facilitate the analysis of waves in anisotropic media.

We begin with the velocity surface, which is the locus of the ends of the vector $\boldsymbol{V} = V\boldsymbol{n}$ drawn from a reference point A. Here V is the phase velocity and $\boldsymbol{n}$ the unit vector in the propagation direction. For isotropic media there will be two velocity surfaces, namely spheres with radii c_α and c_β, for the P-and the S-mode, respectively. For anisotropic media there will be three surfaces, namely one for quasi-longitudinal and two for quasi-shear waves.

Making an inversion of the above surface with respect to the point A we arrive at the slowness surface, $\boldsymbol{L} = \boldsymbol{n}/V$, mentioned in Section 3.4. This possesses the following interesting feature. Denote its radii for $\boldsymbol{n}$ and $\boldsymbol{n} + d\boldsymbol{n}$ as $\boldsymbol{L}$ and $\boldsymbol{L} + d\boldsymbol{L}$, respectively. Then

$$d\boldsymbol{L} = \frac{\partial \boldsymbol{L}}{\partial n_i} dn_i \quad . \tag{3.90}$$

Now we are interested in finding the product $\boldsymbol{V}^e \cdot d\boldsymbol{L}$, where $\boldsymbol{V}^e$ is the energy velocity. Since

$$\frac{\partial L_i}{\partial n_j} = \frac{\partial(n_i/V)}{\partial n_j} = \frac{\delta_{ij}}{V} - \frac{n_i}{V^2}\frac{\partial V}{\partial n_j} \quad ,$$

we get

$$V_i^e \frac{\partial L_i}{\partial n_j} = \left(\frac{V_j^e}{V} - \frac{V_i^e n_i}{V^2}\frac{\partial V}{\partial n_j}\right) \quad .$$

In view of (3.39) and (3.41) and taking into account that $\boldsymbol{V}^e$ equals the group velocity, $\boldsymbol{V}^g$, we arrive at the relation

$$V_i^e \frac{\partial L_i}{\partial n_j} = 0 \quad .$$

On invoking (3.90) we eventually find that the product of interest is zero,

$$\boldsymbol{V}^e \cdot d\boldsymbol{L} = V_i^e \frac{\partial L_i}{\partial n_j} dn_j = 0 \quad . \tag{3.91}$$

Since $d\boldsymbol{L}$ is tangential to the slowness surface, the above relation means that the energy velocity is always normal to this surface, as shown in Figure 3-14.

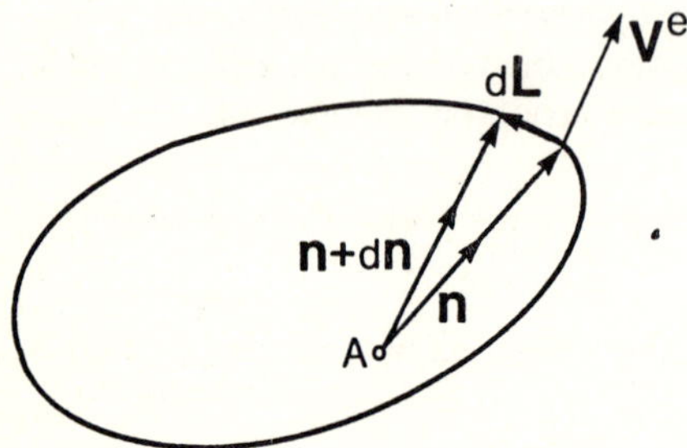

Fig. 3-14 The energy velocity $\boldsymbol{V}^e$ is normal to the slowness surface.

A surface closely related to the previous one is the wave surface, which is the locus of the ends of $\boldsymbol{V}^e$. Thus, the distance between a point of this surface and the origin, A, measured along the given direction, $\boldsymbol{n}$, equals the distance travelled by the energy per unit time. It can be shown that the normal to the wave surface coincides with the propagation direction of a plane wave moving with the coresponding energy velocity, $\boldsymbol{V}^e$.

In fact, since

$$V^e \cdot L = V^e \cdot n/V = 1 \quad ,$$

and in view of (3.39), we get

$$d(V^e \cdot L) = 0 \quad .$$

Taking into account (3.91) we are left with

$$L \cdot dV^e = (n/V)dV^e = 0 \quad , \tag{3.92}$$

which shows that n is normal to dV^e (see Figure 3-15).

Features of the characteristic surfaces indicate a possibility of investigating wave propagation in anisotropic media by graphical means, which seems particularly suitable for engineering practics.

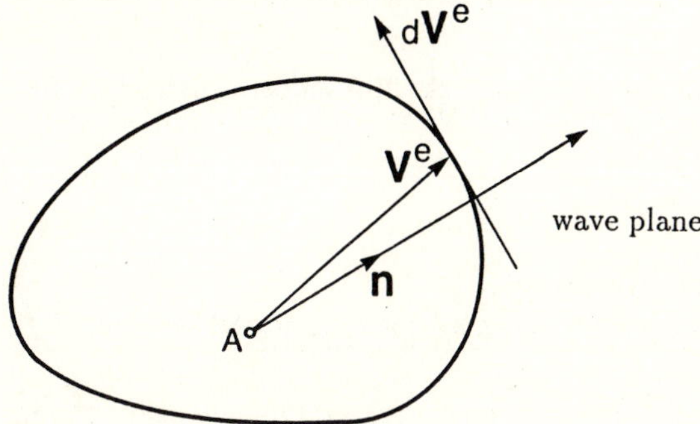

Fig. 3-15 Wave plane is tangential to the wave surface at the corresponding point.

Problems

3-1 Outline the procedure for diagonalization of the tensor $\Gamma_{i\ell}$ and determine the modes polarized in the directions of the principal axes.

3-2 Find the wave velocities and modes along the direction $n_1 = n_2 = n_3$ of a cubic system.

3-3 Figure 3-16 depicts the 45°-cut of a cubic crystal. Waves run in the direction [110]. Find the speed of longitudinal wave and the speeds of two shear waves polarized in the directions $[1\bar{1}0]$ and [001].

3-4 Write down the components of $\tilde{\Gamma}_{i\ell}$ for propagation along the diagonal direction of a cubic piezoelectric system.

3-5 Making use of the results of Problem 3-4, compute the velocities of waves which propagate along the diagonal direction of a cubic piezoelectric system.

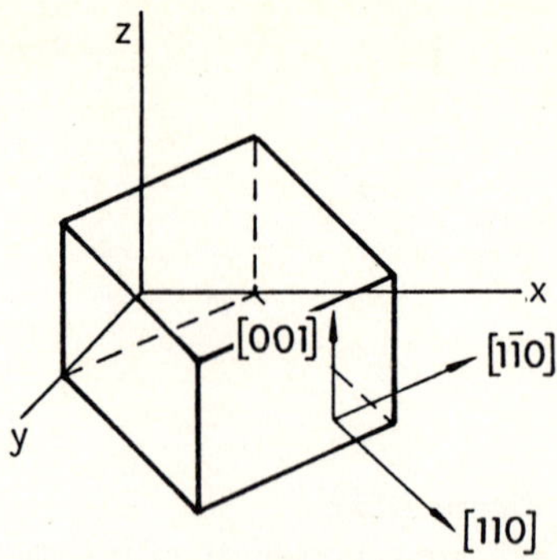

Fig. 3-16 The 45°-cut of a cubic crystal.

3-6 Suggest a method which would take into account the viscoelastic losses in piezoelectric materials.

3-7 How can one find the impulse response of the transducer provided its frequency response is given?

References and Additional Reading

Sections 3.1 – 3.5

Dieulesaint and Royer (1980), Federov (1968), and Musgrave (1970) provide extensive treatments of waves in anisotropic media.

Sections 3.6 – 3.10

Auld (1973), Dieulesaint and Royer (1980), Mariani (1985). See Sachse and Hsu (1979) and Tucker and Rampton (1972) for analysis and technology of transducers. Figs. 3.6, 3.7 and 3.10 are reproduced with the permission of John Wiley & Sons, Ltd. from Dieulesaint and Royer (1980).

Chapter IV

Boundary Effects and Waveguides

The foregoing analysis, which deals with infinite media, may be incomplete or even erroneous when applied to real solids of finite dimensions. In fact, the presence of boundaries gives rise to the phenomena of reflection, surface waves, and resonance.

We begin with presentation of the classic results concerning the reflection of elastic waves at a free boundary and at an interface. Then the Rayleigh waves in isotropic and anisotropic half-spaces are given. In the latter case the consideration of a cubic system enables us to exemplify the theory in a simple way. The further investigation of surface effects deals with Love waves in layered structures and with waves generation by an interdigital transducer, which finds numerous technological applications. Among the omissions is the Lamb problem, which deals with radiation from a point source in a half-space. Considerations of this and related problems would require a substantial increase in the book volume. Nevertheless, these solutions are extensively discussed in literature concerning theoretical foundations of elastic waves.

Behavior of waveguides, such as plates, cylinders, and beams, is treated in a condensed way through exact as well as approximate theories. The spatial decay of disturbances in waveguides is much smaller than in an unbounded medium, which is one of the reasons for their extensive use. The analysis covers free waves and forced response. Among the techniques dealing with the latter, that of a modal superposition is particularly suitable for applied problems.

Material imperfections, such as dislocations or cracks, radiate elastic waves provided the solid is properly loaded. This is frequently referred to as acoustic emission. An analysis of an imperfect rod subjected to a twisting dipole at its ends enables us to illustrate in a simple manner the radical influence, which imperfections may have on a host structure. Another related problem, by which this chapter is completed, deals with radiation from a moving load.

4.1 Reflection and Mode Conversion at a Free Boundary

An elastic half-space serves as a simple model for investigating basic phenomena due to the presence of boundaries. Consider the elastic isotropic half-space defined by $x \geq 0$, $-\infty < y < \infty$, and $-\infty < z < \infty$ and depicted in Figure 4-1, with the z-axis out of the figure plane. When a wave propagating in this half-space strikes the boundary, $x = 0$, reflected disturbances arise. It can be shown that mode conversion occurs upon reflection, namely, either P- or S-wave incident may give rise to both P- and S-waves reflected.

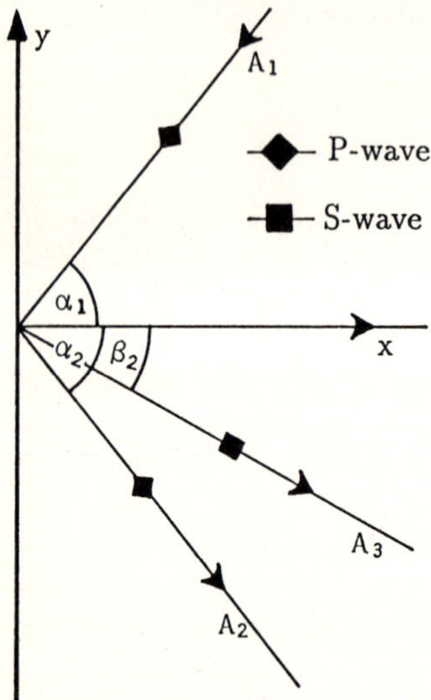

Fig. 4-1 Incident and reflected waves.

Consider, say, the incident dilatational wave travelling in the xy-plane and making the angle α_1 with the x-axis (Figure 4-1). The wave is of infinite beam width, so the problem can be treated as one of plane strain, which implies, among others, $u_z = 0$.

Denoting the incident displacement along the wave normal as u^i we write the harmonic wave as

$$u^i = A_1 e^{i(\phi_1 x + g_1 y + \omega t)} \quad , \tag{4.1}$$

where

$$\phi_1 = k_\alpha \cos \alpha_1, \quad g_1 = k_\alpha \sin \alpha_1, \quad k_\alpha = \omega / c_\alpha \quad .$$

Hence, the disturbance runs in the direction of decreasing x and y until it impinges on the surface $x = 0$.

This surface is taken to be free, which implies the boundary conditions

$$\sigma_{yx}(x = 0, t) = 0 \tag{4.2}$$

$$\sigma_{xx}(x = 0, t) = 0 \quad . \tag{4.3}$$

To satisfy these relations we must stipulate the existence of reflected waves. For the sake of generality, assume that both P- and S-waves are reflected from the boundary. Denoting the displacement for the dilatational wave as u_d^r (along its propagation direction) and that for the shear wave as u_s^r (normal to its propagation direction) we may write

$$u_d^r = A_2 e^{i(-\phi_2 x + g_2 y + \omega t + \delta_1)} \tag{4.4}$$

$$u_s^r = A_3 e^{i(-\phi_3 x + g_3 y + \omega t + \delta_2)} \quad , \tag{4.5}$$

where

$$\phi_2 = k_\alpha \cos\alpha_2, \quad g_2 = k_\alpha \sin\alpha_2$$

$$\phi_3 = k_\beta \cos\beta_2, \quad g_3 = k_\beta \sin\beta_2, \quad k_\beta = \omega/c_\beta \quad ,$$

with the angles α_2, β_2 indicated in Figure 4-1. In writing (4.4) and (4.5) we took into account that this time the disturbances travel in the direction of increasing x. These relations contain six unknowns, the amplitudes, A_2, A_3, the phases, δ_1, δ_2, and the angles α_2, β_2, to be found from the conditions at $x = 0$ given by (4.2) and (4.3).

On computing the total displacements along the axes x and y, u_x and u_y, respectively, we have

$$\begin{aligned} u_x &= u^i \cos\alpha_1 - u_d^r \cos\alpha_2 + u_s^r \sin\beta_2 \\ u_y &= u^i \sin\alpha_1 + u_d^r \sin\alpha_2 + u_s^r \cos\beta_2 \quad . \end{aligned} \tag{4.6}$$

Figure 4-2, which displays decompositions of the waves with respect to the reference frame, is useful for formulating these relations.

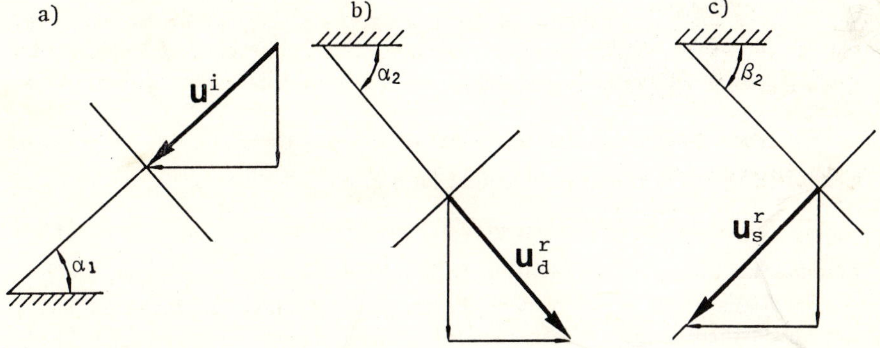

Fig. 4-2 Decomposition of displacements.

Equation (4.2) implies that in the case under consideration

$$u_{x,y} + u_{y,x} = 0 \qquad \text{for} \qquad x = 0 \quad .$$

Making the proper substitutions we get, after some calculations,

$$\begin{aligned}&\frac{A_1}{c_\alpha}\omega \sin 2\alpha_1 \cos(\omega t + g_1 y) - \frac{A_2}{c_\alpha}\omega \sin 2\alpha_2 \cos(\omega t + g_2 y + \delta_1)\\&-\frac{A_3}{c_\beta}\omega \cos 2\beta_2 \cos(\omega t + g_3 y + \delta_2) = 0 \quad .\end{aligned} \tag{4.7}$$

Since this equation must be satisfied for all y and t we may put

$$g_1 = g_2 = g_3 \tag{4.8}$$

and δ_1 and δ_2 equal 0 or π.

Choosing the phases as

$$\delta_1 = \delta_2 = 0 \tag{4.9}$$

we get from (4.7) and (4.8)

$$\sin \alpha_1/c_\alpha = \sin \alpha_2/c_\alpha = \sin \beta_2/c_\beta \tag{4.10}$$

and

$$2(A_1 - A_2)\cos \alpha_1 \sin \beta_2 - A_3 \cos 2\beta_2 = 0 \quad . \tag{4.11}$$

Turning to the remaining condition given by (4.3) and recalling that in the case at hand

$$\sigma_{xx} = (\lambda + 2\mu)u_{x,x} + \lambda u_{y,y} = 0 \qquad \text{for} \qquad x = 0 \quad ,$$

we get, upon substitution of (4.6) and separation of the real part,

$$\begin{aligned}&\frac{A_1}{c_\alpha}\omega(\lambda + 2\mu \cos^2 \alpha_1)\cos(\omega t + g_1 y)\\&\quad + \frac{A_2}{c_\alpha}\omega(\lambda + 2\mu \cos^2 \alpha_2)\cos(\omega t + g_2 y + \delta_1)\\&\quad - \frac{A_3}{c_\beta}\omega(\mu \sin 2\beta_2)\cos(\omega t + g_3 y + \delta_2) = 0 \quad .\end{aligned}$$

In view of (4.8) and (4.9) this implies

$$(A_1 + A_2)(\lambda + 2\mu \cos^2 \alpha_1) - (A_3 c_\alpha \mu \sin 2\beta_2)/c_\beta = 0 \quad ,$$

which can be put in a more convenient form given by

$$(A_1 + A_2)\cos 2\beta_2 \sin \alpha_1 - A_3 \sin \beta_2 \sin 2\beta_2 = 0 \quad . \tag{4.12}$$

Here use has been made of (4.10) and the relation $(\lambda + 2\mu)/\mu = c_\alpha^2/c_\beta^2$.

Equations (4.10), which provide the angles of reflection, are known as Snell's law. Given these angles, the amplitudes, A_2 and A_3, can be found from (4.11) and (4.12), as functions of α_1 and c_α/c_β, which depends solely on Poisson's ratio, ν. The example of such a dependence is given in Figure 4-3 computed for $\nu = 0.25$. We observe that at normal incidence, $\alpha_1 = 0$, no reflected shear waves are generated. Furthermore, $A_2/A_1 = -1$, which means that an incident compression pulse reflects as a tension one and vice versa. In fact, (4.11) and (4.12) are frequency-independent, so all the harmonic components of the pulse obey the same law. Some of the materials exhibit low strength when subjected to tensile stresses and similar phenomenona, when amplified may cause their failure. Figure 4-3 also shows that for $\alpha_1 \simeq 60^\circ$ and $\alpha_1 \simeq 77^\circ$ no reflected dilatational waves occur, the event known as total mode conversion.

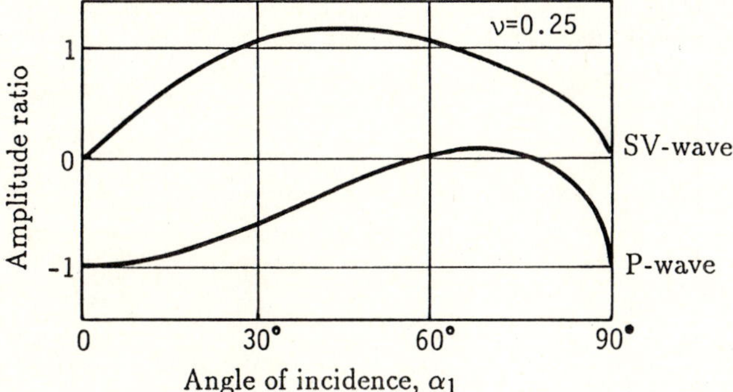

Fig. 4-3 Amplitude ratio for reflected waves.

The reflection of incident shear waves can be analyzed in a similar fashion and is left to exercises. It may be noted, that unlike P- and SV-waves the case of incident SH-waves again presents a simplification, for no mode conversion occurs.

4.2 Reflection at an Interface. Impedance Matching

Reflections at an interface of two half-spaces filled with different elastic materials represent an evident generalization of the previous problem. The analysis follows the same lines and is outlined in the sequel.

There are, this time, four boundary conditions to satisfy, which deal with: i) the normal displacement, ii) the tangential displacement, iii) the normal stress, and iv) the tangential stress. These must be matched across the interface since the media are supposed to be in welded contact. On the other hand, an incident disturbance may give rise to four types of waves: i) two reflected waves propagating back in the first medium, and ii) two refracted waves running forward in the second medium.

The phenomenon at hand is depicted in Figure 4-4. An incident dilatational wave runs in the xy-plane, impinges on the boundary at an angle α_1, and gives rise to the dilatational reflected and refracted waves with angles α_2 and α_3, respectively, as well as to the shear waves with angles, β_2 and β_3. Let A_1 be the incident wave amplitude, whereas A_2, A_3, A_4, and A_5 the wave amplitudes generated upon reflection, as shown in Figure 4-4. Without dwelling on expressions for the waves involved, which are similar to those of the previous section, we turn to the boundary conditions.

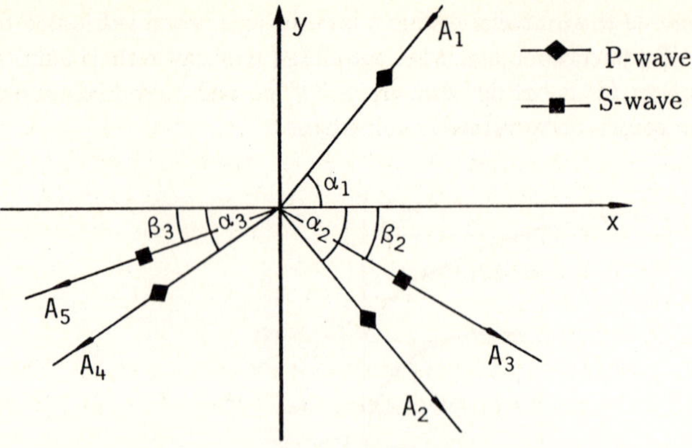

Fig. 4-4 Reflection at an interface.

Matching the normal displacements across $x = 0$ gives

$$(1-\Delta_2)\cos\alpha_1 + \Delta_3\sin\beta_2 - \Delta_4\cos\alpha_3 - \Delta_5\sin\beta_3 = 0, \tag{4.13}$$

while matching the tangential displacements (in the directions y) provides

$$(1+\Delta_2)\sin\alpha_1 + \Delta_3\cos\beta_2 - \Delta_4\sin\alpha_3 + \Delta_5\cos\beta_3 = 0, \tag{4.14}$$

where $\Delta_i = A_i/A_1$.

Next, invoking the continuity of normal and shear stresses across the boundary, we get two additional equations

$$\begin{aligned}(1+\Delta_2)c_{\alpha1}\cos 2\beta_2 - \Delta_3 c_{\beta1}\sin\beta_2 - \Delta_4 c_{\alpha2}(\rho_2/\rho_1)\cos 2\beta_3 \\ -\Delta_5 c_{\beta2}(\rho_2/\rho_1)\sin 2\beta_3 = 0\end{aligned} \tag{4.15}$$

and

$$\begin{aligned}\rho_1 c_{\beta1}^2[(1-\Delta_2)\sin 2\alpha_1 - \Delta_3(c_{\alpha1}/c_{\beta1})\cos 2\beta_2] \\ -\rho_2 c_{\beta2}^2[\Delta_4(c_{\alpha1}/c_{\alpha2})\sin 2\alpha_3 - \Delta_5(c_{\alpha1}/c_{\beta2})\cos 2\beta_3] = 0\ ,\end{aligned} \tag{4.16}$$

where, in notations for the densities and the wave velocities, subscript 1 refers to the first medium and 2 to the second.

The solution to this system imposes Snell's law of reflection, which is given by

$$\begin{aligned}\sin\alpha_1/c_{\alpha1} &= \sin\alpha_2/c_{\alpha1} = \sin\beta_2/c_{\beta1} \\ &= \sin\alpha_3/c_{\alpha2} = \sin\beta_3/c_{\beta2}\ .\end{aligned} \tag{4.17}$$

The case of normal incidence, $\alpha_1 = 0$, is of particular interest. Then (4.13)-(4.17) show that no S-waves occur, $A_3 = A_5 = 0$, while the amplitudes of P-waves are governed by

$$\begin{aligned}\Delta_2 &= A_2/A_1 = (\rho_2 c_{\alpha2} - \rho_1 c_{\alpha1})/(\rho_2 c_{\alpha2} + \rho_1 c_{\alpha1}) \\ \Delta_4 &= A_4/A_1 = 2\rho_1 c_{\alpha1}/(\rho_2 c_{\alpha2} + \rho_1 c_{\alpha1})\ .\end{aligned} \tag{4.18}$$

The product ρc, which is the ratio of the stress to the velocity in an infinite medium, is referred to as the impedance and has been first introduced in Section 1.15. Equations (4.18) indicate that the reflected and refracted amplitudes depend on the impedances involved. In the case of impedance matching, $\rho_1 c_{\alpha 1} = \rho_2 c_{\alpha 2}$, no reflected waves occur, $A_2 = 0$. Hence, the energy of incident wave transfers completely to the second medium. Further, the ratio A_2/A_1 is positive provided that $\rho_2 c_{\alpha 2} > \rho_1 c_{\alpha 1}$. Accordingly, no stress reversal occurs in this case, and, say, a tension pulse reflects as a tension one. However, for a particular case of a free half-space, $\rho_2 c_{\alpha 2} = 0$, one observes that the above effect does take place, as has been noted in the previous section.

As Snell's law indicates limitations must be put on the angle of incidence to insure the existence of the refracted waves described. In fact, from (4.17) we have

$$\sin \alpha_3 = \sin \alpha_1 / (c_{\alpha 1}/c_{\alpha 2}) \leq 1, \qquad (\sin \alpha_1 > 0) \quad .$$

It follows that an incident P-wave may generate a P-wave in the second medium only if

$$\sin \alpha_1 < (c_{\alpha 1}/c_{\alpha 2}) \quad .$$

When this condition is violated, the phenomenon, which is known as total reflection, must be treated in terms of complex rather than real quantities. In this case the analysis shows the occurrence of disturbances which decay exponentially with the distance from the interface.

4.3 Rayleigh Waves in an Isotropic Half-Space

As has been mentioned in Section 4.1, an elastic half-space is apparently the simplest model, which enables us to evaluate effects of a free boundary on the wave propagation. In fact, in this section we shall find elastic waves in a half-space, which are entirely different from those in an infinitely extended medium.

Let the xy-plane be the boundary free of tractions with z positive towards the solid interior (Figure 4-5). Assume the state of plane strain with $u_y = 0$. The displacements u_x and u_z should then be independent of the y-coordinate.

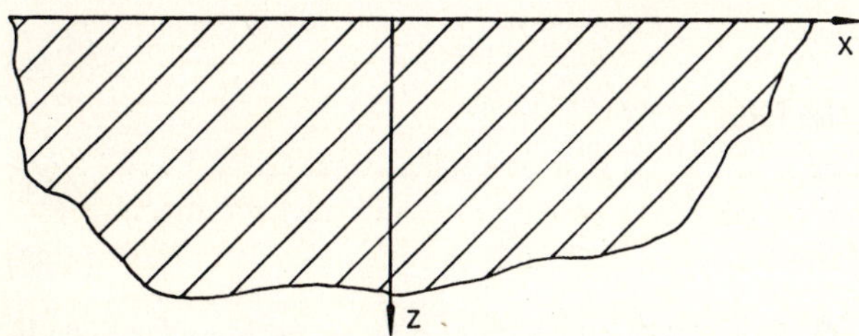

Fig. 4-5 To the theory of Rayleigh waves.

On invoking (2.21) and adapting them to plane strain we arrive at the relations

$$u = u_\alpha + u_\beta$$

$$u_x = \Pi_{,x} + F_{,z}\,, \qquad u_z = \Pi_{,z} - F_{,x} \tag{4.19}$$

where Π and $F = F_y$ are the potentials which provide the dilatational (u_α) and shear (u_β) components, respectively. The potentials are governed by (2.22) and (2.23).

Now we look for waves which run in the x-direction and stay close to the surface $z = 0$ in the sense that their amplitude decays with the depth, z. It is natural to try the following solutions to (2.22) and (2.23)

$$\Pi = \Pi_o(z)e^{i(\omega t - k_R x)} \tag{4.20}$$

$$F = \psi_o(z)e^{i(\omega t - k_R x)} \quad . \tag{4.21}$$

Here $\Pi_o(z)$ and $\psi_o(z)$ are unknown amplitude functions and k_R is as yet unknown wave number, while frequency, ω, is a stipulated value.

On substituting (4.20) and (4.21) into (2.22) and (2.23), respectively, we get

$$\omega^2\Pi_o(z)/c_\alpha^2 = k_R^2\Pi_o(z) - \Pi_o(z)_{,zz} \tag{4.22}$$

and

$$\omega^2\psi_o(z)/c_\beta^2 = k_R^2\psi_o(z) - \psi_o(z)_{,zz} \quad . \tag{4.23}$$

The general solution to (4.22) is

$$\Pi_o(z) = Ce^{-qz} + C_1e^{qz} \quad , \tag{4.24}$$

with

$$q^2 = k_R^2 - \omega^2/c_\alpha^2 \quad . \tag{4.25}$$

Depending on q, one of the terms in (4.24) leads to an increase in the amplitude, when an observer moves towards the interior of the solid. We put $C_1 = 0$, setting, thus, q is real and positive to enforce amplitude decay. In a similar way we obtain the proper solution to (4.23) as

$$\psi_o(z) = De^{-sz} \quad , \tag{4.26}$$

with

$$s^2 = k_R^2 - \omega^2/c_\beta^2 \quad . \tag{4.27}$$

Now the potentials take the form

$$\Pi = Ce^{-qz + i(\omega t - k_R x)} \tag{4.28}$$

$$F = De^{-sz + i(\omega t - k_R x)} \quad , \tag{4.29}$$

where the unknowns, C, D, and k_R, should be found from the boundary conditions at $z = 0$.

Making use of Hooke's law and (4.19) we find

$$\sigma_{zz} = (\lambda + 2\mu)\Pi_{,zz} + \lambda\Pi_{,xx} - 2\mu\psi_{,xz}$$
$$\sigma_{zx} = \mu(2\Pi_{,xz} - \psi_{,xx} + \psi_{,zz}) \quad .$$

Substituting (4.28) and (4.29) we get for $z = 0$

$$\begin{aligned} &C\Big[(\lambda + 2\mu)q^2 - \lambda k_R^2\Big] - 2D\mu i s k_R = 0 \\ &2iqk_R C + (s^2 + k_R^2)D = 0 \quad . \end{aligned} \tag{4.30}$$

The solution to these simultaneous homogeneous equations exists provided the determinant vanishes, which yields the following third equation

$$\Big[(\lambda + 2\mu)q^2 - \lambda k_R\Big](s^2 + k_R^2) - 4\mu q s k_R^2 = 0 \quad .$$

Taking into account (2.7) and introducing the Rayleigh wave velocity, $c_R = \omega/k_R$, this relation can be manipulated into

$$\gamma^6 - 8\gamma^4 + 8(3 - 2/\chi_o^2)\gamma^2 + 16(1/\chi_o^2 - 1) = 0 \quad , \tag{4.31}$$

where

$$\gamma = c_R/c_\beta \tag{4.32}$$

$$\chi_o = c_\alpha/c_\beta = [2(1-\nu)/(1-2\nu)]^{1/2} \quad . \tag{4.33}$$

The wave velocity, c_R, can be found from this algebraic equation, which depends solely on Poisson's ratio, ν. Accordingly, the Rayleigh waves in a purely elastic medium are non-dispersive.

For clearer physical understanding insert (4.28) and (4.29) into (4.19) and separate the real parts to obtain the displacements

$$\begin{aligned} u_x &= Ck_R\Big[e^{-qz} - 2qse^{-sz}/(s^2 + k_R^2)\Big]\sin(\omega t - k_R x) \\ u_z &= Cq\Big[e^{-qz} - 2k_R^2 e^{-sz}/(s^2 + k_R^2)\Big]\cos(\omega t - k_R x) \quad , \end{aligned} \tag{4.34}$$

where D has been eliminated by means of (4.30). The coefficient C remains arbitrary. Since $u_y = 0$, the particle displacement occurs in the sagittal xz-plane. Equations (4.34) indicate an elliptical trajectory regardless of the depth, z. The rate of decay with z depends on the factors q and s, which can be written as

$$\begin{aligned} q^2 &= (1 - \gamma^2/\chi_o^2)k_R^2 \\ s^2 &= (1 - \gamma^2)k_R^2 \quad . \end{aligned} \tag{4.35}$$

Accordingly, the ellipse shape changes and the polarization, which is negative at $z = 0$, may then become positive.

It remains to discuss in some detail (4.31), which provides the wave velocity as a function of Poisson's ratio. It can be shown that for $0 < \nu < 1/2$ one real positive root, which always lies in the interval $0 < \gamma < 1$, does exist. Hence, by virtue of (4.32)

$$c_R < c_\beta < c_\alpha \quad .$$

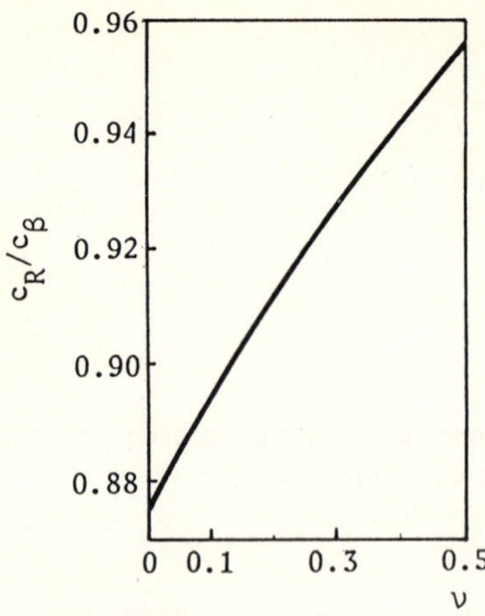

Fig. 4-6 The wave velocity c_R vs. Poisson's ratio.

Figure 4-6 computed from (4.31) shows c_R/c_β as a function of ν.

4.4 Rayleigh Waves in an Anisotropic Half-Space

This section extends the analysis of surface waves to anisotropic materials. Unlike the previous case, this time directions in planes parallel to the free boundary surface are not equivalent and the problem becomes three- dimensional, which renders the computations much more involved.

4.4.1 General Equations

Assume that the half-space occupies the region $x_3 \geq 0$, $-\infty < x_2 < \infty$, $-\infty < x_1 < \infty$, as shown in Figure 4-7, and introduce the propagation direction by vector $\boldsymbol{n}_2 = (n_1,\ n_2,\ 0)$ and a field point by vector $\boldsymbol{r}_2 = (x_1,\ x_2,\ 0)$. Relying on the previous analysis we look for surface waves, the amplitude of which decays with the depth. These may be written in terms of displacements, u_k, which run in the direction parallel to the x_1x_2-plane,

$$u_k = A_k e^{-\chi x_3} e^{i\omega(t - \boldsymbol{n}_2 \cdot \boldsymbol{r}_2 / c_R)}, \quad k = 1, 2, 3 \quad . \tag{4.36}$$

Here the wave velocity, c_R, the decay parameter, χ, and the amplitude factor, A_k, are unknown values, while the direction of propagation, $\boldsymbol{n}_2$, may be considered as given.

Whether such waves may exist should be verified from the standpoint of the equations governing the phenomenon. The equations of motion are

$$c_{ijk\ell} u_{\ell,jk} = \rho \ddot{u}_i \tag{4.37}$$

and the traction-free boundary conditions are

$$\sigma_{i3} = c_{i3k\ell} u_{k,\ell} = 0 \qquad \text{for} \qquad x_3 = 0 \quad , \tag{4.38}$$

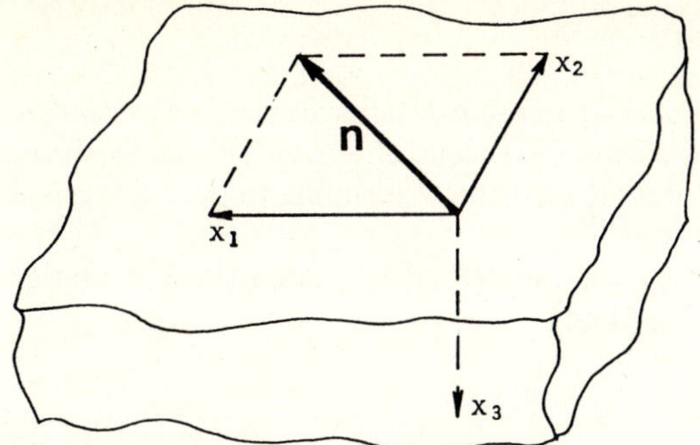

Fig. 4-7 To the theory of Rayleigh waves in anisotropic media.

according to (3.1) and (1.31). Here $c_{ijk\ell}$ are the elastic moduli.

Further calculations may be greatly simplified by introducing vector $\boldsymbol{r} = (x_1, x_2, x_3)$ and vector $\boldsymbol{q} = (q_1, q_2, q_3)$, which is given by

$$q_1 = n_1, \quad q_2 = n_2, \quad q_3 = c_R\chi/(i\omega) \tag{4.39}$$

Then (4.36) takes the standard form

$$u_k = A_k e^{i\omega(t - \boldsymbol{q}\cdot\boldsymbol{r}/c_R)} \tag{4.40}$$

and the treatment becomes similar to that of Section 3.1.

In fact, substitution of (4.40) into (4.37) yields the Christoffel equation

$$(\Gamma_{i\ell} - \rho c_R^2 \delta_{i\ell})A_\ell = 0 \quad , \tag{4.41}$$

where

$$\Gamma_{i\ell} = c_{ijk\ell} q_j q_k \quad . \tag{4.42}$$

The determinant of this system must vanish to ensure the existence of a solution, which gives

$$|\Gamma_{i\ell} - \rho c_R^2 \delta_{i\ell}| = 0 \quad . \tag{4.43}$$

As follows from (4.36), only those roots of this equation, which provide $\mathrm{Re}\,\chi > 0$, may be of interest. Denote these as $q_3^{(\lambda)}$, $\lambda = 1, 2, 3$ since (4.43) is cubic in q_3^2. Each of the roots has the associated displacement vector, $u_k^{(\lambda)}$, emerging from (4.41) and (4.40). Then the general solution is given by the linear combination

$$u_k = \sum_{\lambda=1}^{3} a_\lambda u_k^{(\lambda)} \quad , \tag{4.44}$$

where the coefficients a_λ and the velocity, c_R, must be found from the remaining equation (4.38), which expresses the boundary conditions.

4.4.2 Cubic System

To make this procedure clearer consider in some detail a Rayleigh wave propagation, say, in the [100] direction of the (001) plane of a cubic system (a silicon crystal, for example). In many aspects the treatment is similar to that of Sections 3.1 - 3.3.

Substituting $n_1 = 1$, $n_2 = 0$, $n_3 = q_3$ into (3.13) we obtain the propagation tensor, $\Gamma_{i\ell}$, appearing in (4.41) and (4.42)

$$\begin{aligned}&\Gamma_{11} = c_{11} + c_{44}q_3^2, \quad \Gamma_{12} = \Gamma_{21} = 0,\\ &\Gamma_{13} = \Gamma_{31} = (c_{12} + c_{44})q_3, \quad \Gamma_{22} = c_{44}(1 + q_3^2),\\ &\Gamma_{23} = \Gamma_{32} = 0, \quad \Gamma_{33} = c_{44} + c_{11}q_3^2 \quad .\end{aligned} \tag{4.45}$$

Here use has been made of (1.32) to simplify notations.

Equation (4.43), governing the roots, $q_3^{(\lambda)}$, takes the form

$$(\Gamma_{22} - \rho c_R^2)\left[(\Gamma_{11} - \rho c_R^2)(\Gamma_{33} - \rho c_R^2) - \Gamma_{13}^2\right] = 0 \quad . \tag{4.46}$$

Hence, one of the roots follows from

$$\Gamma_{22} - \rho c_R^2 = 0 \tag{4.47}$$

and the other two from the remaining equation, which is quartic in q_3. Only the roots which imply $\operatorname{Re}\chi > 0$ are of interest.

For example, from (4.47) and (4.45) we get

$$c_{44}(1 + q_3^2) - \rho c_R^2 = 0 \quad .$$

On denoting the relevant root as $q_3^{(2)}$, $(\lambda = 2)$, we obtain

$$\left[q_3^{(2)}\right]^2 = \rho c_R^2 / c_{44} - 1 \quad ,$$

where c_R is as yet unknown.

Since according to (4.39)

$$\chi^{(2)} = i\omega q_3^{(2)} / c_R \quad ,$$

that value of $q_3^{(2)}$, which has a negative imaginary part, would ensure $\operatorname{Re}\chi^{(2)} > 0$. Accordingly, we choose the root given by

$$q_3^{(2)} = -i(1 - \rho c_R^2 / c_{44})^{1/2} \quad , \tag{4.48}$$

with

$$c_R^2 < c_{44} / \rho \quad . \tag{4.49}$$

The roots of the remaining quartic equation are treated in a similar way.

Now turn to the displacements, $u_k^{(\lambda)}$, associated with the above roots. The explicit form of (4.41) is

$$\begin{aligned} \Gamma_{11}A_1^{(\lambda)} + \Gamma_{13}A_3^{(\lambda)} &= \rho c_R^2 A_1^{(\lambda)} \\ \Gamma_{22}A_2^{(\lambda)} &= \rho c_R^2 A_2^{(\lambda)} \\ \Gamma_{31}A_1^{(\lambda)} + \Gamma_{33}A_3^{(\lambda)} &= \rho c_R^2 A_3^{(\lambda)} \quad , \end{aligned}$$

where use has been made of (4.45). It is seen that the displacements associated with the root $q_3^{(2)}$, $(\lambda = 2)$, given by (4.47) and (4.48), are

$$A_1^{(2)} = A_3^{(2)} = 0, \qquad A_2^{(2)} \text{ -- arbitrary}$$

and, hence, in the normalized form

$$u_1^{(2)} = u_3^{(2)} = 0, \quad u_2^{(2)} = 1 \quad . \tag{4.50}$$

In a similar way, substituting $q_3^{(1)}$ and $q_3^{(3)}$, which are the roots of the quartic equation, we may obtain after some algebra

$$\begin{aligned} u_1^{(1)} &= 1, \quad u_2^{(1)} = 0, \quad u_3^{(1)} = \delta_1 \\ u_1^{(3)} &= 1, \quad u_2^{(3)} = 0, \quad u_3^{(3)} = \delta_3 \quad , \end{aligned} \tag{4.51}$$

where

$$\delta_\varepsilon = \frac{\rho c_R^2 - c_{11} - c_{44}\left[q_3^{(\varepsilon)}\right]^2}{c_{12} + c_{44}\left[q_3^{(\varepsilon)}\right]^2}, \qquad \varepsilon = 1, 3 \quad .$$

Constructing the linear combination, (4.44), and incorporating the boundary conditions, (4.38), we get

$$\begin{aligned} &(\delta_1 + q_3^{(1)})a_1 + (\delta_3 + q_3^{(3)})a_3 = 0 \\ &q_3^{(2)}a_2 = 0 \\ &(c_{11}q_3^{(1)}\delta_1 + c_{12})a_1 + (c_{11}q_3^{(3)}\delta_3 + c_{12})a_3 = 0 \quad . \end{aligned} \tag{4.52}$$

Hence, $a_2 = 0$, which implies that the determinant of the remaining two equations must vanish. This may be shown to yield after algebraic manipulations

$$R^2(1 - R) - \left[1 - (c_{12}/c_{11})^2 - R\right]^2 (1 - c_{11}R/c_{44}) = 0 \quad , \tag{4.53}$$

with

$$R = \rho c_R^2 / c_{11} \quad ,$$

which governs the wave velocity, c_R.

The final expressions for the displacements are

$$
\begin{aligned}
u_1 &= \left(a_1\delta_1 e^{-ik_R q_3^{(1)} x_3} + a_3\delta_3 e^{-ik_R q_3^{(3)} x_3}\right) e^{i\omega(t - x_1/c_R)} \\
u_2 &= 0 \\
u_3 &= \left(a_1\delta_1 e^{-ik_R q_3^{(1)} x_3} + a_3\delta_3 e^{-ik_R q_3^{(3)} x_3}\right) e^{i\omega(t - x_1/c_R)} \quad ,
\end{aligned}
\tag{4.54}
$$

where $k_R = \omega/c_R$. The amplitude factors, a_1 and a_3, are related by

$$
a_3/a_1 = -[\delta_1 + q_3^{(1)}]/[\delta_3 + q_3^{(3)}] \tag{4.55}
$$

as seen from (4.52).

4.5 Love Waves

A layered structure may exhibit the behavior which is entirely different from that of a homogeneous medium. One of the typical manifestations of this, known as Love waves, is considered in the sequel. Further investigations of layered bodies will be given in Chapter 5.

Consider a layer of width d, which is supported by a substrate. The layer and the substrate are infinite and made of different elastic materials (Figure 4-8). We shall show that shear surface waves of displacements, $u_y = u_y(x, z)$, parallel to the boundary, may travel in this structure.

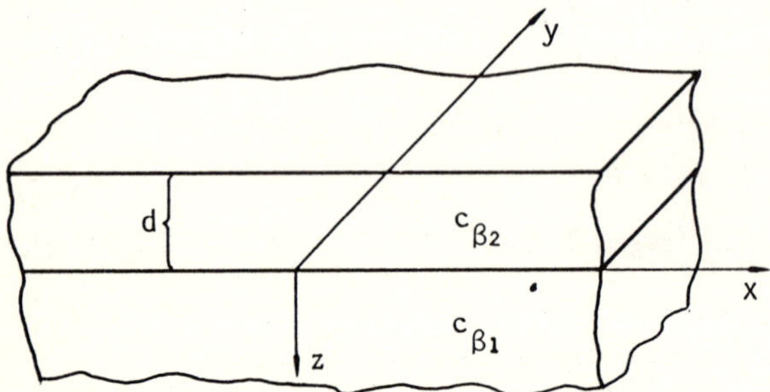

Fig. 4-8 To the theory of Love waves.

To this end, let subscripts 1 and 2 label values related to the substrate and the layer, respectively. Since $u_y = u_y(x, z)$, equations of motion reduce to (2.25), which describe the SH-mode. Adjusting the notations we get

$$
\frac{\partial^2 u_{yi}}{\partial x^2} + \frac{\partial^2 u_{yi}}{\partial z^2} = \frac{1}{c_{\beta i}^2}\frac{\partial^2 u_{yi}}{\partial t^2}, \quad i = 1, 2 \tag{4.56}
$$

where $c_{\beta i}$ is the proper shear wave velocity.

The boundary conditions imply zero tractions at $z = -d$ and continuity of the displacements and stresses at $z = 0$ (welded contact). Hence,

$$u_{y2,z} = 0, \qquad \text{for } z = -d \tag{4.57}$$

$$u_{y1} = u_{y2}, \qquad \text{for } z = 0 \tag{4.58}$$

$$\rho_2 c_{\beta 2}^2 u_{y2,z} = \rho_1 c_{\beta 1}^2 u_{y1,z}, \quad \text{for } z = 0. \tag{4.59}$$

The solutions of interest to (4.56) may be found in the form

$$u_{yi} = f_i(z) e^{i(\omega t - k_L x)}, \quad i = 1,2 \tag{4.60}$$

where the wave number, $k_L = \omega / c_L$, and the functions, $f_i(z)$, are to be specified from the above expressions and the constraint

$$\lim_{z \to \infty} u_{y1} = 0 \quad , \tag{4.61}$$

which imposes a decay with depth, z, typical of the surface waves.

Substitution of (4.60) in (4.56) provides

$$\partial^2 f_1(z)/\partial z^2 - k_L^2 \beta^2 f_1(z) = 0 \tag{4.62}$$

$$\partial^2 f_2(z)/\partial z^2 + k_L^2 \alpha^2 f_2(z) = 0 \quad , \tag{4.63}$$

with

$$\alpha^2 = (c_L/c_{\beta 2})^2 - 1, \qquad \beta^2 = 1 - (c_L/c_{\beta 1})^2 \quad .$$

To ensure stationary vibrations of the layer and, on the other hand, a decay of running waves in the substrate, α and β should be real and positive, which implies

$$c_{\beta 2} < c_L < c_{\beta 1} \quad .$$

Hence, $f_i(z)$ are given by

$$f_1(z) = Ce^{-\beta k_L z} + De^{\beta k_L z} \tag{4.64}$$

$$f_2(z) = A \sin(\alpha k_L z) + B \cos(\alpha k_L z), \tag{4.65}$$

with constant A, B, C, and D.

Equation (4.61) implies $D = 0$, whereas (4.58) and (4.59) give

$$B = C$$

$$A = -C \rho_1 c_{\beta 1}^2 \beta / (\rho_2 c_{\beta 2}^2 \alpha) \quad .$$

Finally, the displacements in the substrate and the layer are found to be

$$u_{y1} = Ce^{-\beta k_L z + i(\omega t - k_L z)} \tag{4.66}$$

$$u_{y2} = C \left[\cos(\alpha k_L z) - \frac{\rho_1 c_{\beta 1}^2 \beta}{\rho_2 c_{\beta 2}^2 \alpha} \sin(\alpha k_L z) \right] e^{i(\omega t - k_L x)} \quad , \tag{4.67}$$

with an arbitrary C.

It remains to find the wave number, k_L, and, consequently, the wave velocity of the Love waves, c_L. This is obtained from (4.57), which, with the help of (4.67), gives

$$\tan(\alpha k_L d) = \rho_1 c_{\beta 1}^2 \beta / (\rho_2 c_{\beta 2}^2 \alpha) \quad , \tag{4.68}$$

which provides c_L as a function of the parameters of the structure and frequency. The Love waves are, thus, of sole shear and are dispersive, unlike the Rayleigh waves.

4.6 Interdigital Transducer

Filters and delay lines may contain interdigital transducers, a simple version of which is depicted in Figure 4-9. It consists of two metal comb shaped electrodes placed on a piezoelectric substrate. An electric field created by the voltage, v, applied to the electrodes, gives rise to dynamic strains in the substrate, which, in turn, launch elastic waves. These contain, among others, the Rayleigh waves, which run in the direction normal to the comb fingers with velocity c_R.

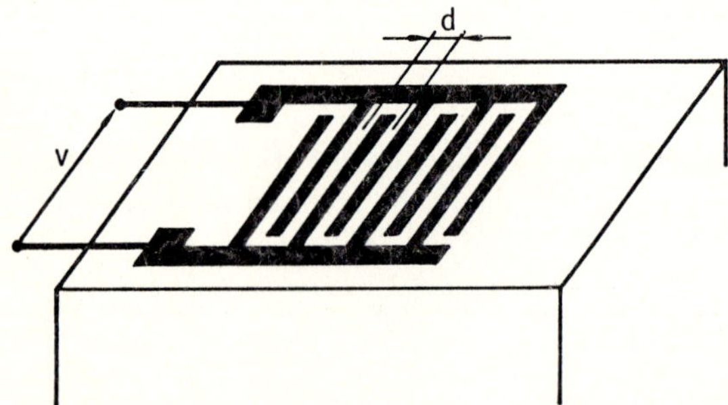

Fig. 4-9 Interdigital transducer.

Since the system is linear it is sufficient to consider the case when applied voltage is harmonic, $v = v_o e^{i\omega t}$. Any stress produced by a finger pair at time t travels along the distance $\lambda_R/2$ during the half period, $T_R/2$. It is desirable that at time $t + T_R/2$ the next pair of fingers will change the voltage sign to generate in-phase stress and insure constructive interference. Hence, the distance d between two neighboring fingers should be equal to half the elastic wavelength

$$d = \lambda_R/2 \quad .$$

The associated frequency, ω_o, is known as synchronous and is given by

$$\omega_o = \pi c_R/d$$

or

$$f_o = c_R/(2d) \quad . \tag{4.69}$$

The exact calculation of the piezoelectric field launched by the interdigital transducer is rather involved. We shall therefore confine the analysis to the heuristic discrete source method, which provides a simple and reasonable approximation.

According to this approach each finger pair is viewed as a discrete ultrasonic source located midway between two fingers. The amplitude assigned to such a source is taken to be proportional to the overlap length of the pair involved, ℓ, with sign depending on the direction of the electric field (Figure 4-10). Radiation is assumed to be uniform along the finger length. Summation of plane waves from all these sources provides the total field. Hence, the approach neglects coupling between elastic and electric quantities as well as diffraction by fingers.

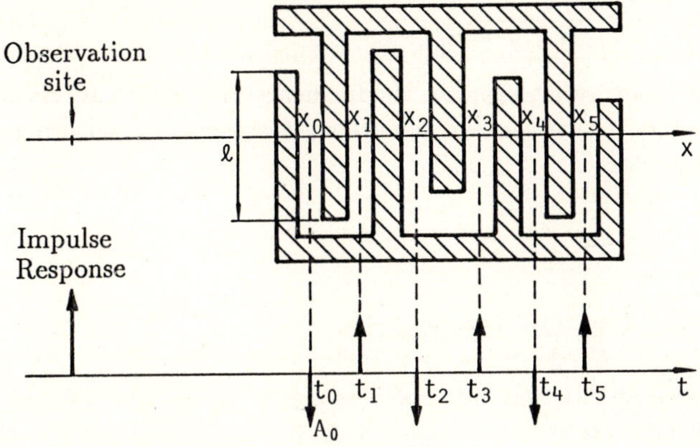

Fig. 4-10 To the discrete source method.

A source located at x_n radiates waves, u_n, travelling, say, in the negative x-direction along the substrate surface

$$u_n = s_n A_n e^{-i\omega(x_n - x)/c_R} e^{i\omega t} \quad ,$$

where, according to the above remarks,

$$s_n = (-1)^n, \qquad A_n = A_o = \text{const}$$

and c_R is the speed of the Rayleigh waves.

The "amplitude" of this wave, B_n, can be written as

$$B_n = s_n A_n e^{-i\omega x_n / c_R} = s_n A_n e^{-i2\pi f t_n} \quad ,$$

with $t_n = (x_o + nd)/c_R$, as Figure 4-10 shows.

On assuming that the transducer has N equally placed fingers, ($N-1$ finger pairs), the total field amplitude at the observation point, 0, or the frequency response, $H(if)$, may be defined as

$$H(if) = \sum_{n=0}^{N-2} B_n = A_o \sum_{n=0}^{N-2} (-1)^n e^{-i2\pi f t_n} \quad .$$

In doing so we neglect similar waves travelling in the positive x-direction since they run away from the detection point.

Taking into account that according to (4.69)

$$t_n = t_o + nd/c_R = t_o + n/(2f_o)$$

we get

$$H(if) = A_o e^{-i2\pi f t_o} \sum_{n=0}^{N-2} e^{-i\pi n(f/f_o - 1)} \quad ,$$

with $t_o = x_o/c_R$. This can be finally written as

$$H(if) = A_o e^{-i2\pi f t_o} \frac{1 - e^{-i(N-1)\pi(f/f_o - 1)}}{1 - e^{-i\pi(f/f_o - 1)}} \quad . \tag{4.70}$$

This simple approximate result provides the frequency response in terms of the transducer parameters, N and f_o, and the driving frequency, f, which may be used for design purposes.

4.7 Waves in a Plate. Cut-Off Frequencies

Many of the structural elements may be viewed as elastic layers. To this end, consider an infinite elastic plate shown in Figure 4-11, which occupies the region $|x| < \infty$, $|y| < \infty$, and $|z| \leq h$. The plate is assumed to be in plane strain and to support waves running in the x-direction. Our purpose is to find the dependence between the wave number, k, and the plate parameters for a given frequency, ω, in other words, the dispersion curve.

The general displacement-potential relations are given by (2.21). They define the total displacement, $\boldsymbol{u}$, as

$$\boldsymbol{u} = \boldsymbol{u}_\alpha + \boldsymbol{u}_\beta = \nabla\Pi + \nabla \times \boldsymbol{F} \quad , \tag{4.71}$$

where Π and $\boldsymbol{F}$ are scalar and vector potentials, respectively. Under plain strain conditions, $u_y = 0$, $\partial()/\partial y = 0$, and Π and $\boldsymbol{F}$, can be functions of x and z only. We therefore invoke (4.19), which is a reduced form of three-dimensional relations,

$$\begin{aligned} u &= u_x = \Pi_{,x} + F_{,z} \\ w &= u_z = \Pi_{,z} - F_{,x} \quad , \end{aligned} \tag{4.72}$$

where $F = F_y$.

The wave potentials, Π and $\boldsymbol{F}$, are to obey (2.22) and (2.23). Stipulating

$$\begin{aligned} \Pi &= f(z) e^{i(kx - \omega t)} \\ F &= g(z) e^{i(kx - \omega t)} \end{aligned} \tag{4.73}$$

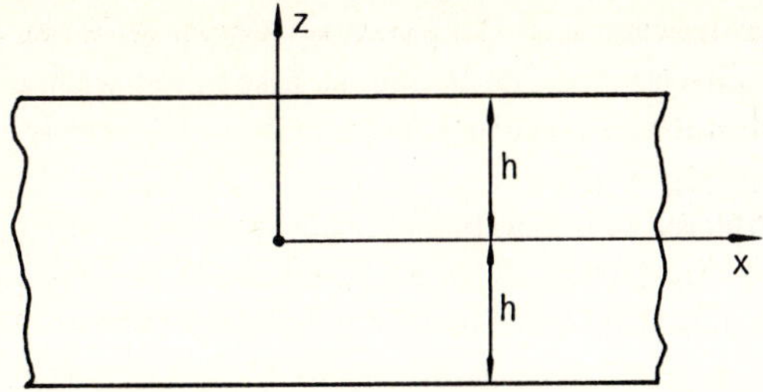

Fig. 4-11 Elastic plate.

and substituting into (2.22) and (2.23) we arrive in both cases at ordinary differential equations of the same type, which can be readily solved to provide

$$\begin{aligned} f(z) &= A \sin \eta_\alpha z + B \cos \eta_\alpha z \\ g(z) &= C \sin \eta_\beta z + D \cos \eta_\beta z \quad , \end{aligned} \tag{4.74}$$

with

$$\eta_\alpha^2 = (\omega / c_\alpha)^2 - k^2 \tag{4.75}$$

$$\eta_\beta^2 = (\omega / c_\beta)^2 - k^2 \tag{4.76}$$

and A, B, C, and D free constants.

Now we write the displacements making use of (4.72)

$$\begin{aligned} u &= [ikf(z) + g'(z)]e^{i(kx-\omega t)} \\ w &= [f'(z) - ikg(z)]e^{i(kx-\omega t)} \quad . \end{aligned} \tag{4.77}$$

Equations (4.77) and (4.74) show that the displacements consist of symmetric and antisymmetric components. The displacement u is symmetric (antisymmetric) with respect to $z = 0$, if it contains only cosines (sines), while w is symmetric (antisymmetric), if it contains only sines (cosines).

The coefficients A, B, C, D and, consequently, the dispersion relation, $k = k(\omega)$, can be specified from the boundary conditions prescribed. This will be considered in the next section. The stresses involved in the boundary conditions at $z = \pm h$ are

$$\begin{aligned} \sigma_{zz} &= \lambda(u_{,x} + w_{,z}) + 2\mu w_{,z} \\ \sigma_{zx} &= \mu(w_{,x} + u_{,z}) \quad . \end{aligned} \tag{4.78}$$

The dispersion relations, $k = k(\omega)$, of elastic waveguides possess usually a complicated branching structure and infinite number of modes. Another typical feature is the presence of cut-off frequencies. To this end, consider (4.73), for example. For a complex wave number, $k = \gamma + i\delta$, the wave term in (4.73) gets the form

$$e^{-\delta x} e^{i(\gamma x - \omega t)} \quad .$$

It is easily seen that for a purely imaginary wave number, $\gamma = 0$, this term does not describe any running disturbance but a standing wave. Note that $\gamma < 0$ might still be appropriate for a disturbance progressing in the opposite direction.

It follows that the case $k = 0$ separates between propagating modes given by $|\,\mathrm{Re}\,k| > 0$ and non-propagating by $|\,\mathrm{Re}\,k| = 0$. The frequency associated with a vanishing wave number, $k = 0$, is referred to as a cut-off frequency, ω_c. Obviously, in this case we get for the phase velocity, $c \to \infty$.

The elastic layer may also be subjected to out-of-plane displacements running in the x-direction. These SH-waves are governed by an equation of the type applied in the theory of Love waves and their investigation is left to exercises.

4.8 Frequency Spectrum of a Plate

A particularly simple analysis applies to the case of the so-called lubricated faces given by

$$w = \sigma_{zx} = 0 \qquad \text{at} \qquad z = \pm h \quad . \tag{4.79}$$

Making use of the second of (4.77) and the second of (4.78) we obtain the following four homogeneous equations

$$\begin{aligned}
&\eta_\alpha \cos\alpha \cdot A - \eta_\alpha \sin\alpha \cdot B - ik\sin\beta \cdot C - ik\cos\beta \cdot D = 0\\
&\eta_\alpha \cos\alpha \cdot A + \eta_\alpha \sin\alpha \cdot B + ik\sin\beta \cdot C - ik\cos\beta \cdot D = 0\\
&i2k\eta_\alpha \cos\alpha \cdot A - i2k\eta_\alpha \sin\alpha \cdot B - (\eta_\beta^2 - k^2)\sin\beta \cdot C\\
&\quad -(\eta_\beta^2 - k^2)\cos\beta \cdot D = 0\\
&i2k\eta_\alpha \cos\alpha \cdot A + i2k\eta_\alpha \sin\alpha \cdot B + (\eta_\beta^2 - k^2)\sin\beta \cdot C\\
&\quad -(\eta_\beta^2 - k^2)\cos\beta \cdot D = 0 \quad ,
\end{aligned} \tag{4.80}$$

where $\alpha = \eta_\alpha h$, $\beta = \eta_\beta h$. The determinant of this system is equal to zero, if

$$\sin\alpha\cos\alpha\sin\beta\cos\beta = 0 \quad , \tag{4.81}$$

which gives four uncoupled non-trivial solutions

$$\begin{aligned}
&\sin\eta_\alpha h = 0, \quad \cos\eta_\alpha h = 0\\
&\sin\eta_\beta h = 0, \quad \cos\eta_\beta h = 0 \quad .
\end{aligned} \tag{4.82}$$

If we choose, say, the first of these relations, then from (4.80) we get

$$A = C = D = 0, \quad \sin\eta_\alpha h = 0, \quad \eta_\alpha = m\pi/(2h), \quad m = 0, 2, 4 \ldots \tag{4.83}$$

while for the second of (4.82) we get

$$B = C = D = 0, \quad \cos\eta_\alpha h = 0, \quad \eta_\alpha = m\pi/(2h), \quad m = 1, 3, 5 \ldots \tag{4.84}$$

Similarly,

$$A = B = D = 0, \quad \sin\eta_\beta h = 0, \quad \eta_\beta = n\pi/(2h), \quad n = 0, 2, 4, 6 \ldots \tag{4.85}$$

$$A = B = C = 0, \quad \cos\eta_\beta h = 0, \quad \eta_\beta = n\pi/(2h), \quad n = 1, 3, 5 \ldots \tag{4.86}$$

Hence, the dilatational modes, represented by the coefficients A and B, and shear modes, represented by C and D, are in this case uncoupled. The values of integers, m and n, specify the so-called modes of propagation, with (4.83) and (4.85) representing symmetric modes (with respect to the midplane) and (4.84) and (4.86) antisymmetric ones, as they have been introduced in the previous sections. Figure 4-12 depicts qualitatively some of these modes for u and w.

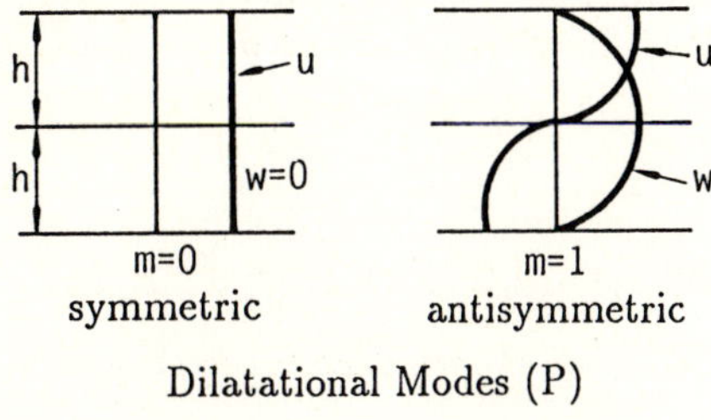

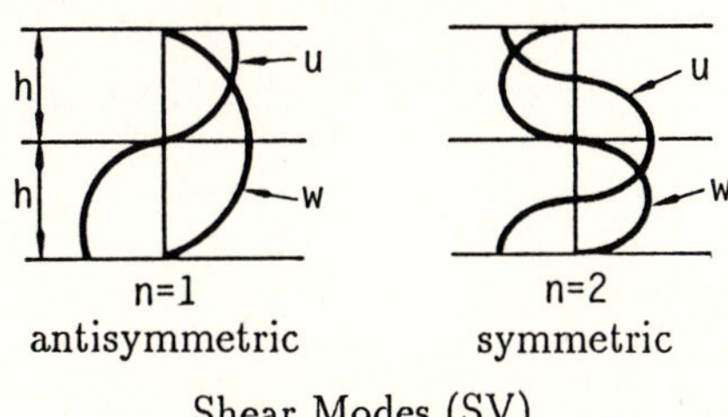

Fig. 4-12 Symmetric and antisymmetric components.

From (4.83), (4.84) and (4.75) we get the dispersion relation for dilatational modes

$$(\omega/c_\alpha)^2 = [m\pi/(2h)]^2 + k^2 \quad ,$$

which can be rewritten in the dimensionless form as

$$\Omega^2 = (c_\alpha/c_\beta)^2(m^2 + \zeta^2) = [\omega 2h/(\pi c_\beta)]^2 \quad , \tag{4.87}$$

where $\zeta = 2hk/\pi$. The entities, Ω and ζ, may be referred to as the dimensionless frequency and the dimensionless wave number, respectively. Equation (4.87) reveals that for a fixed m there is a continuous spectrum of $\omega - k$ pairs, which constitute a branch of the frequency spectrum. This relation may also be written as

$$c^2 = [(m/\zeta)^2 + 1]c_\alpha^2 \quad , \tag{4.88}$$

where use has been made of the relation $k = \omega/c$. Similar results can be obtained for the shear modes too.

It is seen from (4.87) and (4.88) that the dimensionless wave number, ζ, must be real, $\zeta = \gamma$, or purely imaginary, $\zeta = i\delta$. The latter case describes the standing wave, which decays with x, as can be readily seen from (4.73). The cut-off frequencies are given by $\gamma = \delta = 0$.

An example of the frequency spectrum is depicted in Figure 4-13 for $\nu = 1/3$. For $m = 0$ we have $\Omega = (c_\alpha/c_\beta)\gamma$, which gives $c = c_\alpha$ and, hence, represents the dilatational mode of the infinitely extended medium. The anharmonic overtones are given by $m = 1, 2, 3...$ A similar analysis can be carried out for the shear modes, which again shows a typical branching structure of the frequency spectrum and the presence of cut-off frequencies.

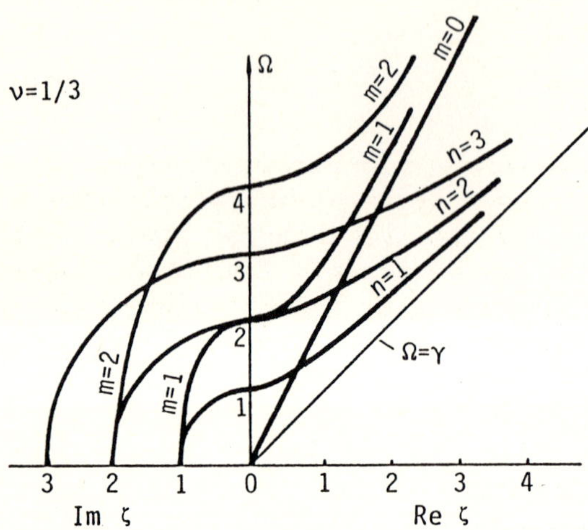

Fig. 4-13 Frequency spectrum for $\nu = 1/3$.

The speed of propagation of a narrow-band wave packet is given by the group velocity, c_g. Applying (1.99) to (4.87) we get

$$c_g/c_\beta = (c_\alpha/c_\beta)^2\gamma/\Omega \quad . \tag{4.89}$$

The boundary conditions of the type treated are somewhat artificial. It is therefore of interest to consider a more realistic case of the traction-free boundaries given by

$$\sigma_{zz} = \sigma_{zx} = 0 \qquad \text{for} \qquad z = \pm h \quad . \tag{4.90}$$

The analysis is similar to the previous one and is summarised below.

Equations (4.77), (4.74) yield the following separate expressions for symmetric modes

$$\begin{aligned} u &= ik\beta \cos\eta_\alpha z + \eta_\beta C \cos\eta_\beta z \\ w &= -\eta_\alpha B \sin\eta_\alpha z - ikC \sin\eta_\beta z \end{aligned} \tag{4.91}$$

and for antisymmetric modes

$$\begin{aligned} u &= ikA \sin\eta_\alpha z - \eta_\beta D \sin\eta_\beta z \\ w &= \eta_\alpha A \cos\eta_\alpha z - ikD \cos\eta_\beta z \quad , \end{aligned} \tag{4.92}$$

where the multiplier $e^{i(kx-\omega t)}$ has been omitted.

The boundary conditions given by (4.90) and (4.78) provide two simultaneous homogeneous equations for symmetric modes (with respect to B and C), and for antisymmetric modes (with respect to A and D). The following relations must hold to ensure the solutions for symmetric modes:

$$\frac{\tan(\eta_\beta h)}{\tan(\eta_\alpha h)} = -\frac{4k^2\eta_\alpha\eta_\beta}{(\eta_\beta^2 - k^2)^2} \tag{4.93}$$

and for antisymmetric modes:

$$\frac{\tan(\eta_\beta h)}{\tan(\eta_\alpha h)} = -\frac{(\eta_\beta^2 - k^2)^2}{4k^2\eta_\alpha\eta_\beta} \quad . \tag{4.94}$$

Equations (4.93) and (4.94) define the Rayleigh-Lamb frequency spectrum, which has the familiar intensively branching structure depending on Poisson's ratio, ν. If frequency is taken real and positive, then the wave number may be either real or complex. The latter case corresponds to waves the amplitude of which decays in the x-direction. A detailed analysis of the frequency spectrum can be found in the literature concerning specifically motions in elastic plates.

4.9 Torsional Waves in a Cylinder

A long cylinder of circular cross section (Figure 4-14) is one of the basic elastic waveguides and its dynamic response, or the so-called Pochhammer waves, is of substantial interest. We shall need the strain-displacement relations in cylindrical coordinates given by

$$\begin{aligned} \varepsilon_{rr} &= u_{r,r}, \quad \varepsilon_{\theta\theta} = (u_{\theta,\theta} + u_r)/r, \quad \varepsilon_{zz} = u_{z,z} \\ 2\varepsilon_{r\theta} &= u_{r,\theta}/r + r(u_\theta/r)_{,r}\,, \quad 2\varepsilon_{z\theta} = u_{z,\theta}/r + u_{\theta,z} \\ 2\varepsilon_{rz} &= u_{r,z} + u_{z,r} \end{aligned} \tag{4.95}$$

and Hooke's law by

$$\begin{aligned} \sigma_{rr} &= \lambda\varepsilon_{ii} + 2\mu\varepsilon_{rr}, \quad \sigma_{\theta\theta} = \lambda\varepsilon_{ii} + 2\mu\varepsilon_{\theta\theta}, \\ \sigma_{zz} &= \lambda\varepsilon_{ii} + 2\mu\varepsilon_{zz}, \\ \sigma_{r\theta} &= 2\mu\varepsilon_{r\theta}, \quad \sigma_{z\theta} = 2\mu\varepsilon_{z\theta}, \quad \sigma_{rz} = 2\mu\varepsilon_{rz} \quad . \end{aligned} \tag{4.96}$$

In the case of axially symmetric torsional waves we may obviously stipulate

$$u_r = u_z = \partial()/\partial\theta = 0, \quad u_\theta = u_\theta(r, z, t) \quad .$$

Then (2.59) show that the wave potentials Π and χ can be taken zero, while the third potential, ψ, provides u_θ,

$$u_\theta = -\psi_{,r} \quad . \tag{4.97}$$

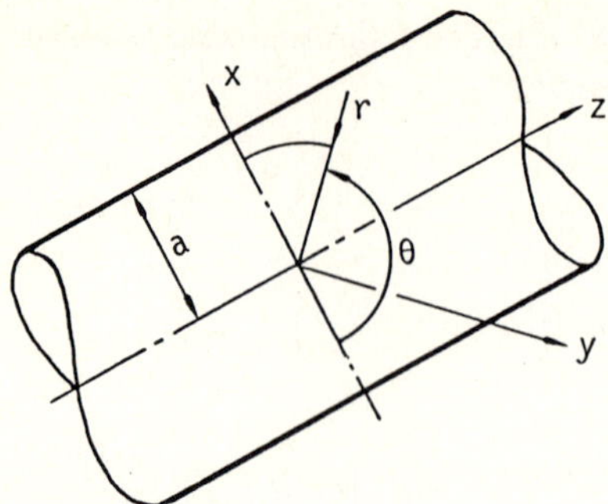

Fig. 4-14 Elastic cylinder.

Invoking (4.95) and (4.96) we obtain the stresses in terms of the potential, ψ. In particular, $\sigma_{r\theta}$, involved in the boundary conditions, is

$$\sigma_{r\theta} = -\mu r \frac{\partial}{\partial r}\left(\frac{\partial \psi}{r \partial r}\right) \quad . \tag{4.98}$$

The potential ψ must satisfy (2.27), which, for harmonic waves, generates the Helmholtz equation (2.44), namely

$$\psi_{,rr} + \psi_{,r}/r + \psi_{,zz} = \psi/c_\beta^2 \quad .$$

Trying the solution given by

$$\psi = A f(r) e^{i(kz-\omega t)}$$

we get the Bessel equation for the amplitude function, $f(r)$,

$$f_{,rr} + f_{,r}/r + \eta_\beta^2 f = 0 \quad , \tag{4.99}$$

where

$$\eta_\beta^2 = (\omega/c_\beta)^2 - k^2 \quad . \tag{4.100}$$

The proper solution to (4.99) is $J_o(\eta_\beta r)$ and, hence,

$$\psi = A J_o(\eta_\beta r) e^{i(kz-\omega t)} \quad ,$$

with η_β given by (4.100), k the unknown wave number, and A the amplitude.

It remains to satisfy the condition on the traction-free surface, $r = a$,

$$\sigma_{r\theta}(r = a) = 0 \quad .$$

Making use of (4.98) we obtain the frequency equation

$$\eta_\beta a J_2(\eta_\beta a) = 0 \quad .$$

This relation indicates that whether $\eta_\beta a = 0$ or $J_2(\eta_\beta a) = 0$. Accordingly, the associated solutions will be investigated separately in what follows.

Taking $\eta_\beta a = 0$, we note first from (4.99) the trivial solution $f(r) = \text{const}$, which corresponds to $u_\theta = 0$. Second, for $\eta_\beta a = 0$, equation (4.99) has the singular solution $f(r) = \log r$, which is physically inadmissible. However, it has also the solution $f(r) = -r^2/2$, which gives

$$\psi = -Ar^2 e^{i(k_o z - \omega t)}/2 \quad , \tag{4.101}$$

with

$$k_o = \omega/c_\beta = k_\beta \quad ,$$

as follows from (4.100). This is the nondispersive fundamental mode.

The other roots, given by $J_2(\eta_\beta a) = 0$, provide the anharmonic overtones. The first three of them are

$$\varepsilon_1 = 5.136, \quad \varepsilon_2 = 8.417, \quad \varepsilon_3 = 11.620 \quad ,$$

where

$$\varepsilon_n = (\eta_\beta a)_n, \quad n = 1, 2, 3...$$

We conclude with the help of (4.97) that the free torsional waves are given either by

$$u_\theta^o = Are^{i(k_o z - \omega t)}, \quad k_o = \omega/c_\beta$$

or by

$$u_{\theta n} = -(\varepsilon_n/a) A J_1(\varepsilon_n r/a) e^{i(k_n z - \omega t)}$$

$$k_n^2 = (\omega/c_\beta)^2 - \varepsilon_n^2/a^2, \quad n = 1, 2, 3...$$

The last expression enables us to derive the phase and the group velocities, associated with each mode, and their dependence on the wavelength or the frequency.

4.10 Longitudinal Waves in a Cylinder

For axially symmetric compressional waves in an infinite elastic rod we first stipulate by symmetry arguments that

$$u_r = u_r(r, z, t), \quad u_z = u_z(r, z, t), \quad u_\theta = 0 \quad . \tag{4.102}$$

Equations (2.59) show that this time we may put $\psi = 0$, while the two other potentials, Π and χ, and the displacements are related by

$$\begin{aligned} u_r &= \frac{\partial \Pi}{\partial r} + \frac{\partial^2 \tilde{\chi}}{\partial r \partial z} \\ u_z &= \frac{\partial \Pi}{\partial z} - \frac{\partial}{r \partial r}\left(r \frac{\partial \tilde{\chi}}{\partial r}\right) \quad , \end{aligned} \tag{4.103}$$

where $\tilde{\chi} = \ell\chi$. The stresses involved in the boundary conditions on a traction-free surface $r = a$ are

$$\begin{aligned} \sigma_{rr} &= \frac{\lambda}{c_\alpha^2}\Pi_{,tt} + 2\mu\left(\frac{\partial^2\Pi}{\partial r^2} + \frac{\partial^3\tilde{\chi}}{\partial r^2\partial z}\right) \\ \sigma_{rz} &= \mu\left\{2\frac{\partial^2\Pi}{\partial r\partial z} + \frac{\partial^3\tilde{\chi}}{\partial r\partial z^2} - \frac{\partial}{\partial r}\left[\frac{\partial}{r\partial r}\left(r\frac{\partial\tilde{\chi}}{\partial r}\right)\right]\right\} \quad , \end{aligned} \tag{4.104}$$

where use has been made of (4.95) and (4.96).

We try the form of solutions similar to that found in the previous section,

$$\begin{aligned} \Pi &= AJ_o(\eta_\alpha r)e^{i(kz-\omega t)} \\ \tilde{\chi} &= BJ_o(\eta_\beta r)e^{i(kz-\omega t)} \quad , \end{aligned} \tag{4.105}$$

where η_β is still given by (4.100), k is as yet unknown wave number, and

$$\eta_\alpha^2 = (\omega/c_\alpha)^2 - k^2 \quad . \tag{4.106}$$

The boundary conditions require the stresses appearing in (4.104) to vanish for $r = a$. Substituting (4.105) in (4.104) we get the two simultaneous equations

$$\begin{aligned} \sigma_{rr} &= \left[-\frac{\lambda\omega^2}{c_\alpha^2}J_o(\eta_\alpha r) + 2\mu\frac{d^2J_o(\eta_\alpha r)}{dr^2}\right]A + i2\mu k\frac{d^2J_o(\eta_\beta r)}{dr^2}B \\ \sigma_{rz} &= i2k\mu\frac{dJ_o(\eta_\alpha r)}{dr}A - \left\{k^2\frac{dJ_o(\eta_\beta r)}{dr} + \frac{d}{dr}\left[\frac{d}{rdr}\left(r\frac{dJ_o(\eta_\beta r)}{dr}\right)\right]\right\}\mu B \quad , \end{aligned}$$

where the multiplier $e^{i(kz-\omega t)}$ has been omitted. On substituting a for r and setting the determinant to zero we obtain after some manipulations the Pochhammer frequency equation

$$(k^2 - \eta_\beta^2)^2\frac{\alpha J_o(\alpha)}{J_1(\alpha)} + 4k^2\eta_\alpha^2\frac{\beta J_o(\beta)}{J_1(\beta)} = 2\eta_\alpha^2(\eta_\beta^2 + k^2) \quad , \tag{4.107}$$

where $\alpha = \eta_\alpha a$ and $\beta = \eta_\beta a$.

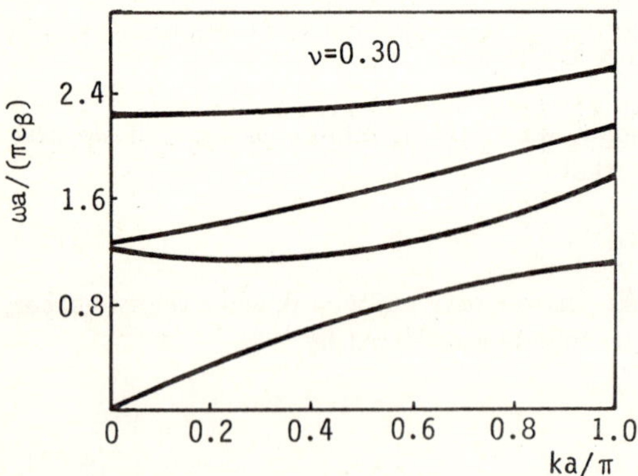

Fig. 4-15 Frequency spectrum for $\nu = 0.3$ (longitudinal waves).

An example of the frequency spectrum following from (4.107) is given in Figure 4-15 for the real dimensionless wave number, $\zeta = ka/\pi$, and the dimensionless frequency, $\Omega = \omega a/(\pi c_\beta)$, $\nu = 0.30$. For $ka \ll 1$, (4.107) may be simplified by making use of expansions of the Bessel functions, which gives

$$\omega a/c_\beta \simeq (E/\mu)^{1/2} ka + 0(ka)^3 \tag{4.108}$$

and provides the well-known elementary approximation

$$c \rightarrow (E/\rho)^{1/2} \quad ,$$

when $ka \rightarrow 0$. Hence, the strength of materials approach is adequate provided that $a/\lambda \rightarrow 0$.

4.11 Flexural Waves in a Cylinder

For flexural waves in an infinite elastic rod all the displacements, u_r, u_θ, and u_z, are functions of the spatial coordinates and time. Hence, all the wave potentials, Π, ψ, and $\tilde{\chi} = \ell\chi$ should be incorporated.

The treatment is similar to the previous problem, but is much more involved. Omitting the Bessel functions of the second kind, Y_n, in the elementary cylindrical wave potential, given by (2.55), to avoid a singularity at $r = 0$, we assume the following expressions

$$\begin{aligned} \Pi &= A_n J_n(\eta_\alpha r) \cos n\Theta e^{i(kz-\omega t)} \\ \psi &= B_n J_n(\eta_\beta r) \sin n\Theta e^{i(kz-\omega t)} \\ \tilde{\chi} &= C_n J_n(\eta_\beta r) \cos n\Theta e^{i(kz-\omega t)} \quad , \end{aligned} \tag{4.109}$$

where $n = 1, 2, 3...$ and η_α and η_β are given by (4.106) and (4.100), respectively. In specifying the angular dependence we took into consideration the symmetry and antisymmetry properties of u_r, u_Θ, and u_z with respect to angle Θ, when the bending occurs in the xz-plane (Figure 4-14). The wave number, $k = \omega/c$, and A_n, B_n, C_n are as yet unknown.

In order to avoid calculations tedious at best, we shall treat only the basic mode, $n = 1$. On inserting (4.109) with $n = 1$ in (2.59) we get after lengthy calculations

$$\begin{aligned} u_r &= R(r) \cos \Theta e^{i(kz-\omega t)} \\ u_\Theta &= \Theta(r) \sin \Theta e^{i(kz-\omega t)} \\ u_z &= Z(r) \cos \Theta e^{i(kz-\omega t)} \quad , \end{aligned} \tag{4.110}$$

where

$$R(r) = \frac{dJ_1(\eta_\alpha r)}{dr} A_1 + \frac{J_1(\eta_\beta r)}{r} B_1 + ik \frac{dJ_1(\eta_\beta r)}{dr} C_1$$

$$\Theta(r) = -\frac{J_1(\eta_\alpha r)}{r}A_1 - \frac{dJ_1(\eta_\beta r)}{dr}B_1 - ik\frac{J_1(\eta_\beta r)}{r}C_1 \tag{4.111}$$

$$Z(r) = ikJ_1(\eta_\alpha r)A_1 + \eta_\beta^2 J_1(\eta_\beta r)C_1$$

Substituting (4.111) and (4.110) into (4.95) and then into (4.96) we obtain the expressions for σ_{rr}, $\sigma_{r\Theta}$ and σ_{rz} involved in the boundary conditions at $r = a$. Three simultaneous homogeneous equations, the determinant of which must vanish, follow. This procedure provides the frequency equation, which governs the wave number, k, and has the following complicated form

$$J_1(\alpha)J_1^2(\beta)(\delta_1 0_\beta^2 + \delta_2 0_\alpha 0_\beta + \delta_3 0_\beta + \delta_4 0_\alpha + \delta_5) = 0 \quad , \tag{4.112}$$

where $\alpha = \eta_\alpha a$, $\beta = \eta_\beta a$, and $0_\gamma = \gamma J_o(\gamma)/J_1(\gamma)$ is the so-called Onoe's function, and

$$\delta_1 = 2(\beta^2 - \zeta^2)^2, \quad \delta_2 = 2\beta^2(5\zeta^2 + \beta^2)$$

$$\delta_3 = \beta^6 - 10\beta^4 - 2\beta^4\zeta^2 + 2\beta^2\zeta^2 + \beta^2\zeta^4 - 4\zeta^4$$

$$\delta_4 = 2\beta^2(2\beta^2\zeta^2 - \beta^2 - 9\zeta^2),$$

$$\delta_5 = \beta^2(-\beta^4 + 8\beta^2 - 2\beta^2\zeta^2 + 8\zeta^2 - \zeta^4) \quad ,$$

with $\zeta = ka$.

This relation generates the branching structure of the frequency spectrum with either real or complex wave number, k. The lowest branch has the asymptotic behavior

$$\omega a/c_\beta = (ka)^2\Big[(1+\nu)/2\Big]^{1/2} + 0(ka)^4 \quad , \tag{4.113}$$

which may be used for computations provided that $ka \ll 1$. Equation (4.113) agrees with the prediction of the elementary theory of bending. The first four modes for the real-valued wave number, k, with $\nu = 0.30$ are shown in Figure 4-16.

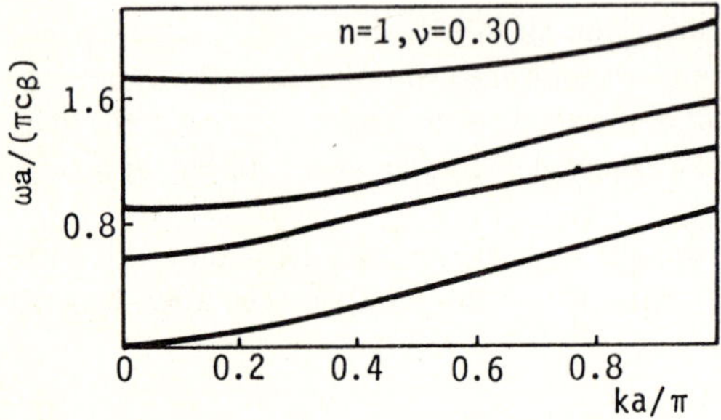

Fig. 4-16 Frequency spectrum for $\nu = 0.3$ (flexural waves).

In conclusion, free harmonic waves in waveguides usually possess a complicated spectrum with real or complex branches and infinite number of modes. This complicates the analysis of transient waves, which may require special techniques. A relevant point to note is that enforcing pure higher modes in practice may be difficult, since a steady input of much energy is needed.

4.12 Timoshenko Beam Theory

The complexity of the exact equations of elastodynamics motivates an extensive use of sundry approximate theories, which may be useful for investigations of transients and/or waveguides with complicated geometry. However, the most simplified models, formulated in the strength of materials, usually fail to predict correctly the high-frequency response as well as a branching structure of the frequency spectrum. A more reasonable approximation follows from models of intermediate level, a typical representative of which is the Timoshenko beam. Hamilton's principle provides a convenient basis for the formulations of the equations governing this model.

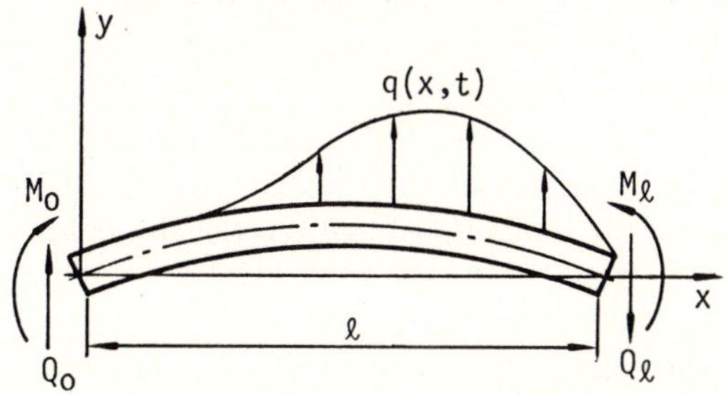

Fig. 4-17 To the theory of Timoshenko beam.

Consider a beam, which vibrates in one of its principle planes and is subject to a lateral load of the intensity $q(x,t)$, to the moment, M, and to the shear force, Q, as shown in Figure 4-17. Denoting the total deflection as $w(x,t)$ and the portion of its slope, which is due solely to bending, as $\alpha(x,t)$, we write the work, A, done by the applied excitation

$$A = \int_0^\ell qw dx - M_o\alpha(x=0,t) + M_\ell\alpha(x=\ell,t) + Q_o w(x=0,t) - Q_\ell w(x=\ell,t) \quad . \tag{4.114}$$

Turning to the strain energy, we imagine the deformations to consist not only of bending but also of shear. Figure 4-18 shows the shear deformations only, which may occur as a result of rotation of line elements tangent to the neutral axis. The elements AB and CD remain vertical, if the shear only takes place. They undergo the total rotation given by

$$\partial w/\partial x = \alpha(x,t) + \beta(x,t) \quad ,$$

where the bending is included too. Obviously, this assumes that cross sections do not remain normal to the neutral axis, contrary to the main assumption of a more elementary approach adopted in the strength of materials.

The strain energy associated solely with bending is

$$\frac{1}{2}\int_0^\ell EI(\partial\alpha/\partial x)^2 dx \quad ,$$

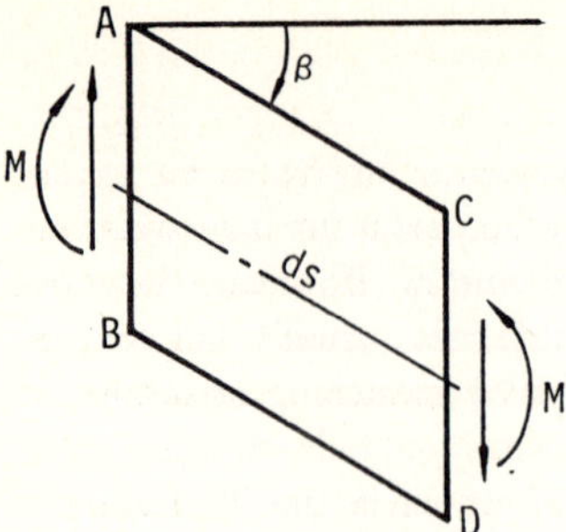

Fig. 4-18 Shear deformations.

while that of the above shear should be a quadratic function of the angle, β,

$$\frac{1}{2}\int_0^\ell k_T\beta^2 dx = \frac{1}{2}\int_0^\ell k_T(\partial w/\partial x - \alpha)^2 dx \quad ,$$

where k_T is a constant to be specified in the sequel.

The kinetic energy, which is due to the translational and rotational motions of an elementary segment of the beam, is

$$\frac{1}{2}\int_0^\ell \left[m(\partial w/\partial t)^2 + I_p(\partial \alpha/\partial t)^2\right] dx \quad ,$$

where m is the mass per unit length and I_p the mass moment of inertia per unit length about the axis normal to the plane of motion and passing through the neutral axis.

Finally, Hamilton's principle, given by (1.64), may be written as

$$\delta \int_{t_o}^{t_1}\int_0^\ell \frac{1}{2}\Big[EI(\partial\alpha/\partial x)^2 + k_T(\partial w/\partial x - \alpha)^2 \\ - m(\partial w/\partial t)^2 - I_p(\partial\alpha/\partial t)^2\Big] dx dt - \delta A = 0 \quad ,$$

with A given by (4.114).

This functional, unlike the elementary theory, contains two unknown functions, $w(x,t)$ and $\alpha(x,t)$, and may thus provide a better model. The functions are governed by the following Euler equations

$$\begin{aligned} &\frac{\partial}{\partial x}\left(EI\frac{\partial\alpha}{\partial x}\right) + k_T\left(\frac{\partial w}{\partial x} - \alpha\right) - I_p\frac{\partial^2\alpha}{\partial t^2} = 0 \\ &m\frac{\partial^2 w}{\partial t^2} - \frac{\partial}{\partial x}\left[k_T\left(\frac{\partial w}{\partial x} - \alpha\right)\right] - q = 0 \quad . \end{aligned} \tag{4.115}$$

with the boundary conditions at each end of the beam given by

either $\delta\alpha = 0$ or $\partial\alpha/\partial x = M/(EI)$

and

either $\delta w = 0$ or $\partial w/\partial x - \alpha = Q/k_T$.

For a homogeneous beam $m = \rho S$ and $I_p = \rho S i_g^2$, where ρ is the mass density, i_g the radius of gyration, and S the cross-sectional area. The constant, k_T, appearing in the expression for the shear strain energy, should be chosen so as to account for a non-uniform distribution of shearing stresses across the beam thickness, which actually takes place. It may also be viewed as a correction factor, which enables one to fit, in one way or another, the exact solution given in Section 4.11. If we adopt the first point of view, then k_T can be expressed as

$$k_T = k_* \mu S$$

The numerical factor, k_*, depends on the shape of the cross-section and may be determined from the static analysis.

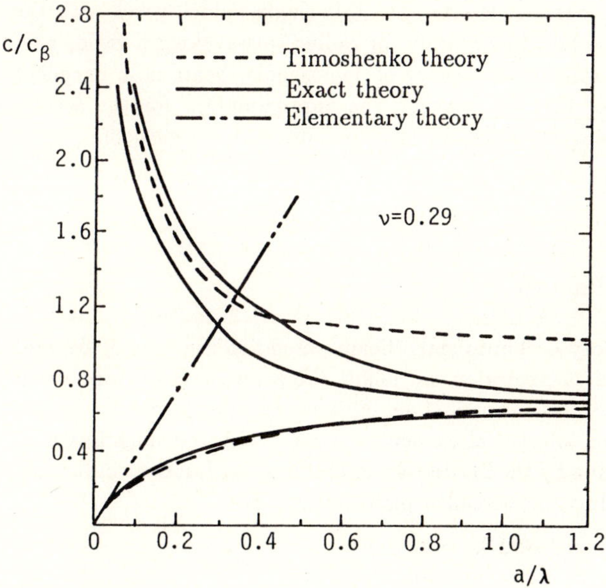

Fig. 4-19 Comparison of theories of flexural waves in a beam.

For prismatic uniform beams with $q = 0$, (4.115) yield the relation

$$\begin{aligned} EI\frac{\partial^4 w}{\partial x^4} + m\frac{\partial^2 w}{\partial t^2} - (I_p + EIm/k_T)\frac{\partial^4 w}{\partial x^2 \partial t^2} \\ + I_p(m/k_T)\frac{\partial^4 w}{\partial t^4} = 0 \quad . \end{aligned} \tag{4.116}$$

On substituting

$$w(x,t) = e^{i(kx-\omega t)}, \quad k = \omega/c = 2\pi/\lambda$$

into (4.116) we get the dispersion relation

$$c^4/(c_b^2 s^2) + c^2(1/c_b^2 + 1/s^2) - c^2/(kc_b i_g)^2 + 1 = 0 \quad ,$$

where

$$c_b = (E/\rho)^{1/2}, \qquad s = (k_*\mu/\rho)^{1/2} \quad .$$

This equation provides obviously two modes (for $c > 0$), which may be anticipated in view of the two functions involved in the description of the deformations. This is an improvement over the elementary theory, which predicts a single mode. However, the model is not yet capable of describing all the details of the wave motion characterized by infinite number of modes, as follows from the exact theory.

Figure 4-19, which shows the results for a beam of circular cross-section, is instructive. It displays the first three modes of the exact theory (solid lines), the two modes of the present theory and the single mode (straight line) following from the elementary approach. It is seen that this single mode agrees with the appropriate mode of the exact theory only for radius to wavelength ratio, a/λ, less than 0.1. The first mode of the theory of Timoshenko beam is in excellent agreement with the exact theory. However, the approximation for the second mode may not be accurate, in particular, for waves of short wavelength.

4.13 Mindlin Plate Theory

Ideas underlying the theory of Timoshenko beam extend to the case of flexural waves in a plate. In the presentation we follow, however, a more formalized procedure.

Figure 4-20 shows the differential element of a plate with the thickness h. It suffers bending and twisting by the moments and by the shear forces as indicated. It is convenient to introduce the so-called plate stresses by

$$(M_{xx}, M_{yy}, M_{xy}) = \int_{-h/2}^{h/2} (\sigma_{xx}, \sigma_{yy}, \sigma_{xy}) z \, dz \tag{4.117}$$

and

$$(Q_{xz}, Q_{yz}) = \int_{-h/2}^{h/2} (\sigma_{xz}, \sigma_{yz}) \, dz \quad , \tag{4.118}$$

where the correspondence among the entities should be clear from these notations. Next, define the plate strains by the similar relations

$$(R_{xx}, R_{yy}, R_{xy}) = (12/h^3) \int_{-h/2}^{h/2} (\varepsilon_{xx}, \varepsilon_{yy}, \varepsilon_{xy}) z \, dz \tag{4.119}$$

and

$$(R_{xz}, R_{yz}) = (2/h) \int_{-h/2}^{h/2} (\varepsilon_{xz}, \varepsilon_{yz}) \, dz \quad . \tag{4.120}$$

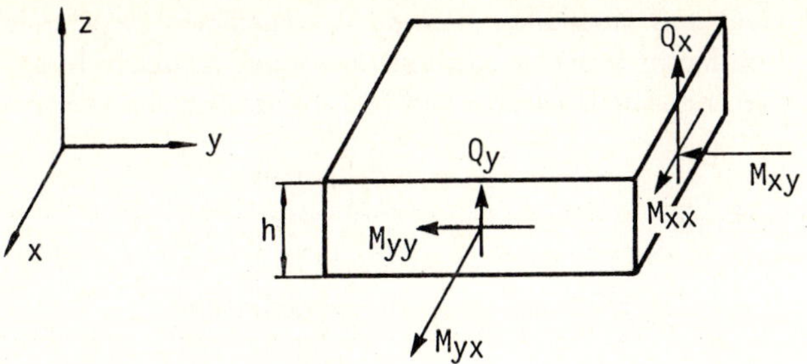

Fig. 4-20 Element of a plate.

4.13.1 Constitutive Law

To find interrelations among the plate stresses and strains we first write ε_{zz} as follows:

$$\varepsilon_{zz} = \left[\sigma_{zz} - \lambda(\varepsilon_{xx} + \varepsilon_{yy})\right]/(\lambda + 2\mu) \quad .$$

Then the remaining equations of Hooke's law may be put in the form

$$\begin{aligned} \sigma_{xx} &= E(\varepsilon_{xx} + \nu\varepsilon_{yy})/(1-\nu^2) + \nu\sigma_{zz}/(1-\nu) \\ \sigma_{yy} &= E(\varepsilon_{yy} + \nu\varepsilon_{xx})/(1-\nu^2) + \nu\sigma_{zz}/(1-\nu) \\ \sigma_{xy} &= 2\mu\varepsilon_{xy}, \quad \sigma_{yz} = 2\mu\varepsilon_{yz}, \quad \sigma_{xz} = 2\mu\varepsilon_{xz} \quad , \end{aligned} \tag{4.121}$$

where the Lame constants, λ and μ, have been replaced by E and ν. Multiplying (4.121) by z and integrating over the thickness of the plate, and then neglecting the term containing σ_{zz}, we get for M_{xx}, M_{yy}, and M_{yx}

$$\begin{aligned} M_{xx} &= D(R_{xx} + \nu R_{yy}), \quad M_{yy} = D(R_{yy} + \nu R_{xx}) \\ M_{yx} &= (1-\nu)DR_{yx} \quad , \end{aligned} \tag{4.122}$$

with

$$D = Eh^3/[12(1-\nu^2)] \quad .$$

Omission of the integral contribution due to σ_{zz} in (4.122), as well as the whole procedure, follow the classical theory of plates. Now we depart from this theory by introducing a shear effect

$$\begin{aligned} Q_{xz} &= 2\mu' \int_{-h/2}^{h/2} \varepsilon_{xz} dx = \mu' h R_{xz} \\ Q_{yz} &= 2\mu' \int_{-h/2}^{h/2} \varepsilon_{yz} dz = \mu' h R_{yz} \quad , \end{aligned} \tag{4.123}$$

where

$$\mu' = k_M \mu \quad . \tag{4.124}$$

The shear stiffness coefficient may thus differ from μ. The coefficient, k_M, may be thought of as a correction factor to be specified in the sequel, similarly to the theory of Timoshenko beam. Equations (4.122) and (4.123) are the constitutive equations of interest.

4.13.2 Equations of Motions

Turning to the strain-displacement relations we recall the definition

$$\varepsilon_{ij} = (u_{i,j} + u_{j,i})/2$$

to obtain

$$\int_{-h/2}^{h/2} (\varepsilon_{xx}, \varepsilon_{yy}, 2\varepsilon_{xy}) z dz = \int_{-h/2}^{h/2} (\partial u/\partial x, \partial v/\partial y, \partial u/\partial y + \partial v/\partial x) z dz \tag{4.125}$$

and

$$2\int_{-h/2}^{h/2} (\varepsilon_{xz}, \varepsilon_{yz}) dz = \int_{-h/2}^{h/2} (\partial u/\partial z + \partial w/\partial x, \partial v/\partial z + \partial w/\partial y) dz \quad , \tag{4.126}$$

where u, v, and w are the displacements along the directions x, y, and z, respectively. In the classical plate theory we put

$$u = z\psi_x(x,y,t), \quad v = z\psi_y(x,y,t), \quad w = w(x,y,t) \quad , \tag{4.127}$$

with

$$\partial w/\partial x = -\psi_x, \quad \partial w/\partial y = -\psi_y \quad . \tag{4.128}$$

In the present approach we release constraints expressed by (4.128) and in the sequel view ψ_x, ψ_y, and w as three independent functions to be specified from the equations of motion, $\sigma_{ij,j} = \rho\ddot{u}_i$, the adaption of which to the case at hand is our next goal.

To this end, multiply the equation of motion given by

$$\sigma_{xx,x} + \sigma_{xy,y} + \sigma_{xz,z} = \rho\ddot{u}$$

by z and integrate across the thickness to get

$$\frac{\partial}{\partial x} M_{xx} + \frac{\partial}{\partial y} M_{xy} + \frac{\partial}{\partial z} \int_{-h/2}^{h/2} \frac{\partial \sigma_{xz}}{\partial z} z dz = \rho \int_{-h/2}^{h/2} \ddot{u} z dz \quad . \tag{4.129}$$

The integral in the left-hand side provides

$$\int_{-h/2}^{h/2} \frac{\partial \sigma_{xz}}{\partial z} z dz = z\sigma_{xz} \Big|_{-h/2}^{h/2} - \int_{-h/2}^{h/2} \sigma_{xz} dz = -Q_{xz} \quad ,$$

since $\sigma_{xz} = 0$ on the boundaries. The integral in the right-hand side can be put in the form

$$\rho \int_{-h/2}^{h/2} \ddot{u} z dz = \rho \psi_{x,tt} h^3/12 \quad ,$$

where use has been made of (4.127). Hence, (4.129) reads

$$M_{xx,x} + M_{xy,y} - Q_{xz} = \rho h^3 \psi_{x,tt}/12 \quad . \tag{4.130}$$

Similarly

$$M_{xy,x} + M_{yy,y} - Q_{yz} = \rho h^3 \psi_{y,tt}/12 \quad . \tag{4.131}$$

The remaining equation of motion gives

$$\frac{\partial}{\partial x} Q_{xz} + \frac{\partial}{\partial y} Q_{yz} + \int_{-h/2}^{h/2} \frac{\partial \sigma_{zz}}{\partial z} dz = -\rho h w_{,tt}$$

If the boundaries are subjected to the tractions

$$\sigma_{zz}(z = h/2, x, y, t) = -q_1(x, y, t)$$
$$\sigma_{zz}(z = -h/2, x, y, t) = -q_2(x, y, t) \quad ,$$

then evaluation of the integral term eventually provides

$$Q_{xz,x} + Q_{yz,y} + q_2 - q_1 = \rho h w_{,tt} \quad , \tag{4.132}$$

which completes the derivation.

With the help of (4.125) and (4.126) the plate equations of motion, (4.130) - (4.132), may be written as

$$D\Big[(1-\nu)\nabla^2\psi_x + (1+\nu)\psi_{,x}\Big]/2 - \mu' h(\psi_x + w_{,x})$$
$$= \rho h^3 \psi_{x,tt}/12$$
$$D\Big[(1-\nu)\nabla^2\psi_y + (1+\nu)\psi_{,y}\Big]/2 - \mu' h(\psi_y + w_{,y})$$
$$= \rho h^3 \psi_{y,tt}/12 \tag{4.133}$$
$$\mu' h(\nabla^2 w + \psi) + q = \rho h w_{,tt} \quad ,$$

where

$$q = q_2 - q_1$$
$$\psi = \psi_{x,x} + \psi_{y,y} \quad .$$

Equations (4.133) govern three independent functions of displacements, ψ_x, ψ_y, and w, and still contain the unspecified coefficient, $\mu' = k_M \mu$, to be found in what follows. A relevant point to note is that the applied procedure, unlike that of the previous section, has not furnished the boundary conditions, which may be deduced by energy considerations.

4.13.3 Dispersion Relation and Specification of the Shear Factor

Equations (4.133) can be transformed to a single expression for w, which is convenient for investigations of free harmonic waves. To this end, differentiate the first two equations and then eliminate ψ to get

$$\left(\nabla^2-\frac{\rho}{\mu'}\frac{\partial^2}{\partial t^2}\right)\left(D\nabla^2-\frac{\rho h^3}{12}\frac{\partial^2}{\partial t^2}\right)w+\rho h\frac{\partial^2 w}{\partial t^2}$$

$$=\left(1-\frac{D\nabla^2}{\mu' h}+\frac{\rho h^2}{12\mu'}\frac{\partial}{\partial t^2}\right)q \quad . \tag{4.134}$$

On letting $q=0$, which may correspond to the traction-free boundaries, and substituting

$$w=e^{i(kx-\omega t)}, \quad k=\omega/c$$

into (4.134) we arrive at the dispersion relation

$$k^2\left[1-c^2/(k_M^2 c_\beta^2)\right](c_N^2/c^2-1)=12/h^2 \quad , \tag{4.135}$$

where

$$c_N^2=E/[\rho(1-\nu^2)] \quad .$$

It remains to specify the coefficient, k_M, which governs μ'. One of the possibilities is to use the geometric limit of (4.135). Then

$$c=k_M c_\beta \qquad \text{as} \qquad \omega\to\infty \quad .$$

On the other hand, the exact theory given in Section 4.11 may be shown to yield

$$c=c_R \qquad \text{as} \qquad \omega\to\infty \quad ,$$

where c_R is the Rayleigh surface wave speed. Accordingly, we prescribe

$$k_M=c_R/c_\beta \quad . \tag{4.136}$$

Computations indicate that this choice leads to a good agreement between the above approach and the exact theory for the full frequency interval, $0\le\omega<\infty$, (see Figure 4-21).

4.14 Normal Modes of Waveguides

Free harmonic waves, like those dealt with in foregoing sections, are instructive as indicators of the dynamic properties of waveguides. Furthermore, they may be used as building blocks for investigations of arbitrary forced motions. The same roles are played by free vibrations of finite waveguides. Here we present quite a general procedure for finding the so-called normal modes of elastic waveguides, closely related to free vibrations. This will then be extended to the case of a forced response in the following section.

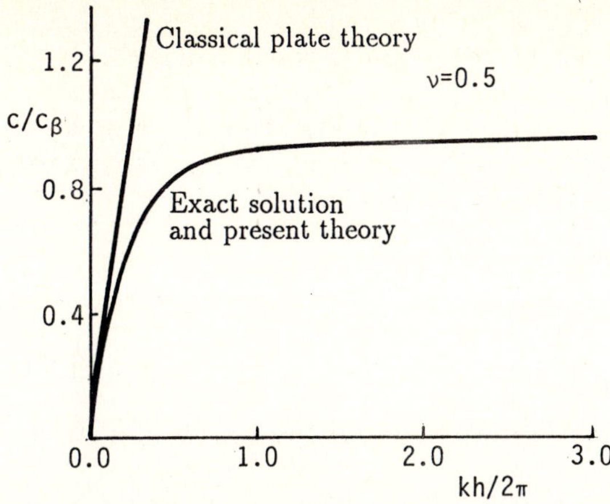

Fig. 4-21 Comparison of theories of flexural waves in a plate.

Consider a typical problem of elastodynamics: find the displacements $u_i(x,t)$ in the domain V bounded by the surface, S, which satisfy for $t > 0$ the following homogeneous boundary conditions

$$\begin{aligned} u_i(x,t) &= 0 \qquad \text{on } S_1 \\ \sigma_{ij}(x,t)n_j &= 0 \qquad \text{on } S_2 \quad , \end{aligned} \tag{4.137}$$

with

$$S = S_1 + S_2$$

and the initial conditions given by

$$\begin{aligned} u_i(x,t) &= u_i^{(0)}(x) \\ \dot{u}_i(x,t) &= \dot{u}_i^{(0)}(x) \quad . \end{aligned} \tag{4.138}$$

Here n is a unit normal to the surface, S.

Unfortunately, now we have no choice but to derive first auxiliary relations of quite a formal nature. We introduce the two systems of the body forces, $f_i^{(m)}$ and $f_i^{(n)}$, and, respectively, the associated displacements, $u_i^{(m)}$ and $u_i^{(n)}$. These functions obey the equations of motion, (2.117) and (2.118), which enables us to construct the formal expression

$$\int_V \left[\left(L_{i\ell}u_\ell^{(m)} + f_i^{(m)}\right)u_i^{(n)} - \left(L_{i\ell}u_\ell^{(n)} + f_i^{(n)}\right)u_i^{(m)}\right]dV = 0 \quad , \tag{4.139}$$

with the operator, $L_{i\ell}$, given by (2.117). On incorporating Hooke's law we get

$$L_{i\ell}u_\ell = \sigma_{ij,j} - \rho\ddot{u}_i$$

and (4.139) transforms to

$$\int_V \Big\{ \Big[\sigma_{ij,j}^{(m)} u_i^{(n)} - \sigma_{ij,j}^{(n)} u_i^{(m)}\Big] + \Big[f_i^{(m)} u_i^{(n)} - f_i^{(n)} u_i^{(m)}\Big] + \rho\Big[\ddot{u}_i^{(n)} u_i^{(m)} - \ddot{u}_i^{(m)} u_i^{(n)}\Big]\Big\} dV = 0 \quad . \tag{4.140}$$

Next, rewrite the first bracketed term as follows:

$$\sigma_{ij,j}^{(m)} u_i^{(n)} - \sigma_{ij,j}^{(n)} u_i^{(m)} = (\sigma_{ij}^{(m)} u_i^{(n)})_{,j} - \Big(\sigma_{ij}^{(n)} u_i^{(m)}\Big)_{,j} - \sigma_{ij}^{(m)} u_{i,j}^{(n)} + \sigma_{ij}^{(n)} u_{i,j}^{(m)} \tag{4.141}$$

and get, with the help of (1.42),

$$\begin{aligned}\sigma_{ij}^{(m)} u_{i,j}^{(n)} &= \Big[\lambda\delta_{ij}\bar{\theta}^{(m)} + \mu\Big(u_{i,j}^{(m)} + u_{j,i}^{(m)}\Big)\Big] u_{i,j}^{(n)} \\ &= \lambda\bar{\theta}^{(m)}\bar{\theta}^{(n)} + \mu u_{i,j}^{(m)} u_{i,j}^{(n)} + \mu u_{j,i}^{(n)} u_{i,j}^{(m)} \quad .\end{aligned} \tag{4.142}$$

Similarly,

$$\sigma_{ij}^{(n)} u_{i,j}^{(m)} = \lambda\bar{\theta}^{(n)}\bar{\theta}^{(m)} + \mu u_{i,j}^{(n)} u_{i,j}^{(m)} + \mu u_{j,i}^{(m)} u_{i,j}^{(n)} \tag{4.143}$$

The last terms in (4.142) and (4.143) are equal, as can be shown by simple interchange of indices. This indicates that the entire expressions are equal too, as their symmetry indeed suggests,

$$\sigma_{ij}^{(m)} u_{i,j}^{(n)} = \sigma_{ij}^{(n)} u_{i,j}^{(m)} \quad . \tag{4.144}$$

This equation is a manifestation of the so-called reciprocal relations.

Returning to (4.141) we note that in view of (4.144) only the first two terms in its right-hand side servive. By the divergence theorem we get for the first bracketed term in (4.140)

$$\begin{aligned}&\int_V \Big[\sigma_{ij,j}^{(m)} u_i^{(n)} - \sigma_{ij,j}^{(n)} u_i^{(m)}\Big] dV \\ &= \int_S \Big[\sigma_{ij}^{(m)} n_j u_i^{(n)} - \sigma_{ij}^{(n)} n_j u_i^{(m)}\Big] ds = 0 \quad ,\end{aligned} \tag{4.145}$$

where use has been made of (4.137). Thus, (4.140) finally reduces to

$$\int_V \rho\Big(\ddot{u}_i^{(n)} u_i^{(m)} - \ddot{u}_i^{(m)} u_i^{(n)}\Big) dV = 0 \tag{4.146}$$

provided the body forces are absent.

Now we return to the physically meaningful case of free vibrations and assume

$$u_i(x,t) = U_i(x)e^{i\omega t} \tag{4.147}$$

Substituting this into the previous equation we get the important orthogonality relation

$$\int_V \rho(x) U_i^{(m)}(x) U_i^{(n)}(x) dV = 0 \quad , \tag{4.148}$$

which holds if $\omega_n^2 \neq \omega_m^2$. Note that this is valid for inhomogeneous waveguides too.

Next, inserting (4.147) into (2.117) and (2.118) with the body forces neglected, $f_i = 0$, we eventually get the equation governing the normal modes, $U_i(x)$,

$$\mu U_{i,jj} + (\lambda + \mu) U_{j,ji} + \rho\omega^2 U_i = 0 \quad . \tag{4.149}$$

This equation is subject to the boundary conditions

$$\begin{aligned} &U_i = 0 && \text{on } S_1 \\ &\left[\lambda\delta_{ij} U_{\ell,\ell} + \mu(U_{i,j} + U_{j,i})\right] n_j = 0 && \text{on } S_2 \quad , \end{aligned} \tag{4.150}$$

stemming from (4.137) and (4.147). The entities, $U^{(n)}$ and ω_n, are also known as the eigenfunctions and the eigenvalues, respectively.

Equations (4.149) and (4.150) specify the normal mode, $U_i(x)$, $i = 1, 2, 3$, within a constant multiplier, which may be found by imposing, for example,

$$\int_V \rho(x) \left[U_i^{(n)}(x) U_i^{(n)}(x)\right] dV = 1 \quad . \tag{4.151}$$

Equations (4.149) – (4.151) usually define a unique and complete set of functions, which describe free vibrations of an elastic waveguide.

4.15 Forced Motions via Modal Superpositions

Now we consider a more general case, assuming in place of (4.137) and (4.138), the nonhomogeneous boundary conditions

$$\begin{aligned} &u_i(x,t) = p_i(x,t) && \text{on } S_1 \\ &\sigma_{ij}(x,t) n_j = g_i(x,t) && \text{on } S_2 \end{aligned} \tag{4.152}$$

and recovering the presence of the body forces, $f_i(x,t)$. We thus deal with forced vibrations of a waveguide.

Making use of linearity of the problem we represent the unknown solution as

$$u_i(x,t) = \tilde{u}_i(x,t) + \sum_{m=1}^{\infty} U_i^{(m)}(x) q^{(m)}(t) \quad . \tag{4.153}$$

Here $q^{(m)}(t)$ are usually called the generalized coordinates, while the matching function $\tilde{u}_i$ satisfies the equation

$$\mu\tilde{u}_{i,jj} + (\lambda + \mu)\tilde{u}_{j,ji} + f_i = 0 \tag{4.154}$$

and the boundary conditions

$$\begin{aligned} &\tilde{u}_i = p_i && \text{on } S_1 \\ &\tilde{\sigma}_{ij} n_j = g_i && \text{on } S_2 \quad , \end{aligned} \tag{4.155}$$

with the obvious relation between $\tilde{\sigma}_{ij}$ and $\tilde{u}_{ij}$. It is seen that the displacements, u_i, satisfy the boundary conditions, (4.152), and may provide the solution upon a proper specification of the generalized coordinates, $q^{(m)}(t)$. To this end, substitute (4.153) in (2.117) and (2.118) to obtain, with the help of (4.149),

$$\sum_{n=1}^{\infty}\left[q_{tt}^{(n)} + \omega_n^2 q^{(n)}\right]\rho U_i^{(n)} = -\rho\tilde{u}_{i,tt} \tag{4.156}$$

A more convenient form is arrived at by multiplying this expression by $U_i^{(n)}$ and integrating over V. This yields

$$q_{,tt}^{(n)} + \omega_n^2 q^{(n)} = Q_{,tt}^{(n)} \quad , \tag{4.157}$$

with

$$Q^{(n)} = -\int_V \rho\tilde{u}_i U_i^{(n)} dv \quad . \tag{4.158}$$

Equation (4.157) has the solution

$$\begin{aligned} q^{(n)}(t) &= q^{(n)}(0)\cos(\omega_n t) + \dot{q}(n)(0)\sin(\omega_n t)/\omega_n \\ &+ \int_0^t Q_{,tt}^{(n)}(\tau)\sin[\omega_n(t-\tau)]d\tau/\omega_n \quad . \end{aligned} \tag{4.159}$$

Therefore, our next purpose is to derive convenient expressions for the forcing term, $Q^{(n)}(t)$, and for the initial values, $q^{(n)}(0)$ and $\dot{q}^{(n)}(0)$.

With the help of (4.149) we get

$$\sigma_{ij,j}^{(n)} = \mu U_{i,jj}^{(n)} + (\lambda+\mu)U_{j,ji}^{(n)} = -\rho\omega_n^2 U_i^{(n)} \quad ,$$

which, in view of (4.158), gives

$$\omega_n^2 Q^{(n)}(t) = \int_{S_1} \sigma_{ij} n_j p_i ds - \int_V \sigma_{ij}^{(n)} \tilde{u}_{i,j} dV \quad ,$$

where use has been made of the boundary conditions and the divergence theorem. Next, because of (4.144)

$$\sigma_{ij}^{(n)}\tilde{u}_{i,j} = \tilde{\sigma}_{ij} U_{i,j}^{(n)} \quad ,$$

and the previous relation acquires the following physically meaningful form

$$Q^{(n)}(t) = \left[-\int_V f_i U_i^{(n)} dv + \int_{S_1} \sigma_{ij}^{(n)} n_j p_i ds - \int_{S_2} U_i^{(n)} g_i ds\right]/\omega_n^2 \tag{4.160}$$

which expresses $Q^{(n)}(t)$ directly in terms of the work done by the body and surface forces.

It remains to deduce the initial conditions for the generalized coordinates, $q^{(n)}(t)$. Making use of (4.153) we get

$$u_i(x,0) = \tilde{u}_i(x,0) + \sum_{n=1}^{\infty} U_i^{(n)}(x) q^{(n)}(0) \quad .$$

On multiplying this equation by $\rho U_i^{(n)}$, and integrating over V, we get

$$q^{(n)}(0) = \int_V \left[\rho u_i(x,0)U_i^{(n)} - \rho\tilde{u}_i(x,0)U_i^{(n)}\right]dV \quad ,$$

which, in view of (4.158), yields

$$q^{(n)}(0) = \int_V \rho u_i(x,0)U_i^{(n)}dV + Q^{(n)}(0) \quad . \tag{4.161}$$

Similarly,

$$\dot{q}^{(n)}(0) = \int_V \rho \dot{u}_i(x,0)U_i^{(n)}dV + \dot{Q}^{(n)}(0) \quad , \tag{4.162}$$

which completes the derivation.

In conclusion, the solution to the problem of forced motion of a waveguide has been obtained in terms of the normal modes, $U_i^{(n)}(x)$, and the generalized coordinates, $q^{(n)}(t)$, given by (4.149) and (4.159), respectively. The forcing term and the initial values involved are defined by (4.160)-(4.162)

This approach applies under fairly general conditions, which makes it convenient for investigating the transient response of complicated waveguides. It may also be used in the frameworks of approximated theories, for example, the theory of Mindlin plate. In fact, a particular form of the governing equations does not affect its basic steps, provided these equations describe a linear elastic system. In the following section we apply it for analysis of an extensional forced response of an elastic layer.

4.15.1 Forced Response of a Layer

Consider the example of inhomogeneous layer, $0 \le x \le h$, $-\infty < y < \infty$, $-\infty < z < \infty$, which is subject to a variable body force, $f(t)$, and to the boundary displacements

$$\begin{aligned} u_x(0,t) &= p(t) \\ u_x(h,t) &= g(t) \quad . \end{aligned} \tag{4.163}$$

The initial conditions are

$$\begin{aligned} u(x,0) &= u_o(x) \\ \dot{u}(x,0) &= \dot{u}_o(x) \quad . \end{aligned} \tag{4.164}$$

In this case we naturally set $u_x = u(x)$, $u_y = u_z = 0$ and write

$$u(x,t) = \tilde{u}(x,t) + \sum_{n=1}^{\infty} U^{(n)}(x)q^{(n)}(t) \tag{4.165}$$

and the following relations for $\tilde{u}(x,t)$:

$$(\lambda + 2\mu)\tilde{u}_{,xx} + f = 0, \quad \tilde{u}(0,t) = p(t), \quad \tilde{u}(h,t) = g(t) \quad , \tag{4.166}$$

where use has been made of (4.153) – (4.155).

The normal modes, $U^{(n)}(x)$, are governed by (4.149) and (4.150), which yield

$$\begin{aligned} &(\lambda + 2\mu)U^{(n)}_{,xx} + \rho\omega_n^2 U^{(n)} = 0 \\ &U^{(n)}(0) = U^{(n)}(h) = 0 \quad . \end{aligned} \tag{4.167}$$

These modes should also be normalized by the condition

$$\int_0^h \rho[U^{(n)}]^2 dx = 1 \quad . \tag{4.168}$$

The initial values of $q(t)$ follow from (4.161) and (4.162)

$$\begin{aligned} q^{(n)}(0) &= \int_0^h \rho u_o U^{(n)} dx + Q^{(n)}(0) \\ \dot{q}^{(n)}(0) &= \int_0^h \rho \dot{u}_o U^{(n)} dx + \dot{Q}^{(n)}(0) \quad . \end{aligned} \tag{4.169}$$

The generalized coordinates, $q^{(n)}(t)$, are now given by (4.159).

To arrive at explicit results we specify the following functions

$$f = u_o = \dot{u}_o = p(t) = 0, \quad \rho = \text{constant} \quad ,$$

while $g(t) \neq 0$. Equations (4.167) and (4.168) provide

$$\begin{aligned} \omega_n &= n\pi[(\lambda + 2\mu)/\rho]^{1/2}/h \\ U^{(n)}(x) &= [2/(\rho h)]^{1/2} \sin(k_n x) \quad , \end{aligned} \tag{4.170}$$

with $k_n = n\pi/h$, $n = 1, 2 ...$ The matching function, $\tilde{u}$, follows from (4.166) as

$$\tilde{u}(x,t) = xg(t)/h$$

and the forcing term, $Q^{(n)}$, from (4.160) as

$$Q^{(n)}(t) = (-1)^n (2\rho h)^{1/2} g(t)/(n\pi) \quad .$$

On invoking also (4.159), (4.170) and (4.165) we eventually find the displacement

$$\begin{aligned} u(x,t) = xg(t)/h + 2\sum_{n=1}^{\infty}[(-1)^n/(n\pi)] \\ \Big\{q(0)\cos(\omega_n t) + \dot{q}(0)\sin(\omega_n t)/\omega_n \\ + \int_0^t \frac{d^2 g(\tau)}{d\tau^2} \sin[\omega_n(t-\tau)]d\tau/\omega_n\Big\} \sin(k_n x) \quad . \end{aligned} \tag{4.171}$$

Techniques based on modal superpositions usually exhibit a satisfactory convergence. However, no general estimates have been obtained yet. The number of modes needed for a reasonable accuracy may vary from 1 to 10^3 or more, depending on a particular problem.

4.16 Acoustic Emission in a Rod

Elastic radiation from imperfections or cracks in stressed finite solids is often referred to as acoustic emission. A resort to the analysis given, for example, in Section 2.16 and dealing with infinite media may lead to erroneous conclusions, since the boundedness of a host structure may have a profound influence on this phenomenon. The following problem provides apparently the simplest possible illustration.

Consider a long circular cylinder with radius a, the mass density, ρ, and the shear modulus, μ, (Figure 4-22). It contains a screw dislocation located at the center and is subject to two twisting moments at the ends as shown in the figure. The moments, $M^d = \pm M_o e^{-i\omega t}$, are always 180° out of phase with one another, so the resultant is zero. Somewhat loosely speaking, this loading may be referred to as a twisting dipole. The forced response of the cylinder must satisfy the equations of motions, the boundary conditions at $r = a$, the singularity conditions for $r \to 0$, and must comply with the given moments, M^d.

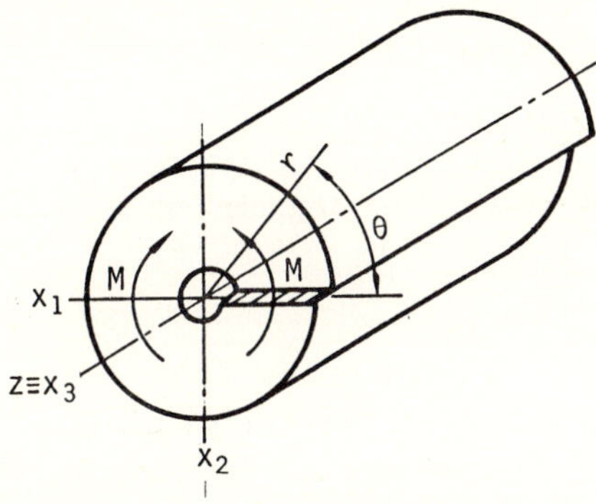

Fig. 4-22 Imperfect cylinder subjected to a twisting dipole.

The solution may be found in the class of SH-waves. We first assume that under the applied moments the dislocation executes harmonic vibrations in the x_1-direction and its position is given by

$$x_1 = de^{-i\omega t}, \quad x_2 = 0 \tag{4.172}$$

with the amplitude, d, which is as yet unknown. This gives rise to shear waves running back and forth inside the cylinder.

Splitting up the displacements in the static, u_i^s, and the dynamic, u_i^d, parts and invoking the results of Section 2.16, we get the following singularity at the dislocation core

$$\begin{aligned} \lim_{r\to 0} u_3^d &= \frac{b\sin\theta}{2\pi r} de^{-i\omega t} \\ \lim_{r\to 0} u_2^d &= \lim_{r\to 0} u_3^d = 0 \quad , \end{aligned} \tag{4.173}$$

where b is the Burgers vector.

Since the static displacements, u_i^s, given by (2.150) produce no stresses at $r = a$, the boundary conditions are

$$\sigma_{rr}^d(r = a, \theta, z, t) = \sigma_{r\theta}^d(r = a, \theta, z, t) = 0$$

and

$$\sigma_{rz}^d(r = a, \theta, z, t) = 0 \quad . \tag{4.174}$$

On setting $u_r = u_\theta = 0$ and $u_z^d = u_z(r, \theta, t)$, the equations of motion are reduced to

$$u_{z,rr}^d + u_{z,r}^d/r + u_{z,\theta\theta}^d/r^2 = \ddot{u}_z^d \rho/\mu \quad .$$

Results of Section 2.5 indicate that a solution of standing waves may be given by

$$u_z^d = \sum_{n=1}^{\infty} \Big[A_n J_n(kr) + B_n Y_n(kr)\Big] \sin n\theta e^{-i\omega t}, \tag{4.175}$$

with

$$k = \omega[\rho/\mu]^{1/2} \quad .$$

Substitution of (4.175) in (4.173) and use of a small argument expansion of $Y_n(kr)$ yield

$$A_n = B_n = 0 \qquad \text{for} \qquad n \neq 1$$

and

$$B_1 = -ibdk/4 \quad .$$

On the other hand, (4.174) provides

$$A_1 J_1'(ka) + B_1 Y_1'(ka) = 0$$

and the forced response of the cylinder is found to be

$$u_z = idbk\Big\{\frac{J_1(kr)[Y_1(ka) - kaY_2(ka)]}{[J_1(ka) - kaJ_2(ka)]} - Y_1(kr)\Big\} \sin\theta/4e^{-i\omega t} \quad , \tag{4.176}$$

where use has been made of recurrence formulae.

It remains to specify the amplitude, d. To this end, calculate the twisting moment

$$M^d = \int_{-\pi/2}^{\pi/2} \int_b^a \sigma_{z\theta} r^2 dr d\theta \quad .$$

Note that the lower limit of integration in the radial direction is taken as b. In fact, in the proximity of the dislocation Hooke's law breaks down for the singularity presence. This proximity should be of the order of the magnitude of the displacement giving rise to the dislocation, that is b.

In view of (4.95), (4.96), we get

$$M^d = \mu \int_{-\pi/2}^{\pi/2} \int_b^a u^d_{z,\theta} r\, dr\, d\theta$$

On substituting (4.176) for u_z, we specify the amplitude, d,

$$d = 2M_o/(ibk\mu Q) \quad ,$$

with

$$Q = \int_b^a \Big\{ \big[Y_1'(ka)/J_1'(ka) \big] J_1(kr) - Y_1(kr) \Big\} r\, dr \quad .$$

The vibrations of the outer surface, $r = a$, are easily observable and are therefore of particular interest. Making use of recurrence formulae of Bessel functions, we find from (4.176) the following simple expression:

$$u^d_z = H(k, a, \theta) d e^{-i\omega t} \quad ,$$

where H is given by

$$H(k, a, \theta) = ib \sin\theta / [2\pi a J_1'(ka)] \quad .$$

Thus, if the amplitude, d, is taken prescribed, the resonances occur at the frequencies associated with the roots of the equation

$$J_1'(ka) = 0 \quad .$$

For example, the first three of them are

$$(ka)_1 = 1.841, \qquad (ka)_2 = 5.441, \qquad \text{and} \qquad (ka)_3 = 8.536 \quad .$$

This problem shows that an imperfection may dramatically change the resonance properties of a host structure and thereby its forced response. In fact, a perfect elastic cylinder has no resonances at the above frequencies.

4.17 Radiation from a Moving Load

Moving loads generate wave propagation even though they may be of a constant magnitude and direction. They represent thus an additional "sort" of wave sources. To illustrate the phenomenon, consider an infinite beam on an elastic foundation, which is subject to a force moving with a constant velocity, v. The geometry of the problem is shown in Figure 4-23. The stiffness of the cross-section

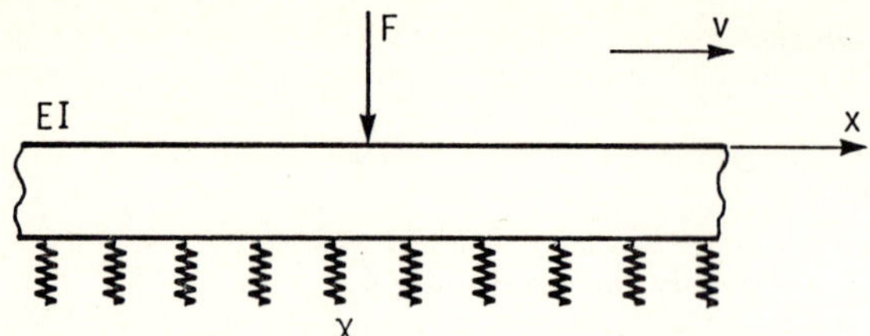

Fig. 4-23 Beam on elastic foundation.

of the beam is given by EI, and its density per unit length by m_o, while the elastic modulus of the foundation by χ.

The equation of motion, which follows from the elementary theory of an elastic beam, is

$$EIw_{,xxxx} + m_o w_{,tt} + \chi w = f \quad , \tag{4.177}$$

with $w(x,t)$ denoting the beam deflection and $f(x,t)$ the intensity of the external load. The terms in the left-hand side of (4.177) describe the elastic force, the inertial force, and the reaction of the foundation, respectively. This equation may be rewritten in a more convenient form

$$w_{,xxxx} + 4\varepsilon_1^2 w_{,tt} + 4\varepsilon_2^4 w = f/(EI) \quad , \tag{4.178}$$

where

$$\varepsilon_1^2 = m_o/(4EI), \quad \varepsilon_2^4 = \chi/(4EI) \quad .$$

For the reasons to become clear in the sequel, we first investigate free waves running along the beam, which are given by

$$w \simeq e^{i2\pi(x-ct)/\lambda} \quad , \tag{4.179}$$

with λ denoting the wavelength and c the phase velocity. Substitution of (4.179) and $f = 0$ in (4.178) yields the dispersion relation

$$c = \left[\varepsilon_2^4 \lambda^2/\pi^2 + (2\pi/\lambda)^2\right]^{1/2}/(2\varepsilon_1) \quad . \tag{4.180}$$

Hence, the phase velocity depends on the wavelength. It follows from (4.180) that c is a minimum when

$$\lambda = \sqrt{2}\pi/\varepsilon_2 = \lambda_o \quad .$$

The associated value of c is referred to as the minimal phase velocity

$$c(\lambda_o) = (4\chi EI/m_o^2)^{1/4} = c_{\min} \quad . \tag{4.181}$$

Having established the dispersive properties of the system we return to the problem posed previously. This time, the forcing term in (4.178) may be written as

$$f(x - vt)/(EI)$$

and the assumed stationary solution as

$$w = w(x - vt) \quad ,$$

where v is the velocity of the load, f.

Introducing the variable $z = x - vt$ we put (4.178) in the form

$$w_{,zzzz} + 4\varepsilon_1^2 v^2 w_{,zz} + 4\varepsilon_2^4 w = f(z)/(EI) \quad , \tag{4.182}$$

with the characteristic equation

$$p^4 + 4\varepsilon_1^2 v^2 p^2 + 4\varepsilon_2^4 = 0 \quad . \tag{4.183}$$

The roots of this equation and, consequently, the solution depend on its coefficients. We therefore confine our further analysis to the case $v < \varepsilon_2/\varepsilon_1$.

Under the above condition the roots of (4.183) are given by

$$p = \pm\alpha\pm\beta i \quad ,$$

with

$$\alpha = (\varepsilon_2^2 - v^2\varepsilon_1^2)^{1/2}, \quad \beta = (\varepsilon_2^2 + v^2\varepsilon_1^2)^{1/2} \tag{4.184}$$

and the solution to (4.182) is

$$\begin{aligned} w = {} & e^{\alpha z}(D_1 \cos\beta z + D_2 \sin\beta z) \\ & + e^{-\alpha z}(D_3 \cos\beta z + D_4 \sin\beta z) + \psi(z) \quad . \end{aligned} \tag{4.185}$$

Here $\psi(z)$ denotes a particular solution which depends on the forcing term and which can be, in fact, omitted, as the following consideration shows. The point $z = 0$ ($x = vt$) indicates the position of the force. Then, for $z > 0$ as well as for $z < 0$ there is no external force and the solution is given by (4.185) with $\psi = 0$.

Rejecting the terms, which show spatial growth of waves, we get from (4.185)

$$w = e^{\alpha z}(D_1 \cos\beta z + D_2 \sin\beta z), \quad \text{for } z < 0$$

$$w = e^{-\alpha z}(D_3 \cos\beta z + D_4 \sin\beta z), \quad \text{for } z > 0 \quad .$$

Three of the above coefficients may be found by matching the above solutions and its derivatives up to the second order at $z = 0$. For the third derivative we additionally get

$$[w_{,zzz}] = F/(EI) \quad ,$$

with F denoting the concentrated force, shown in Figure 4-23, and the bracketed term being the value of discontinuity at $z = 0$. This yields

$$w = -Fe^{\mp\alpha z}[\cos\beta z \pm \alpha \sin(\beta z)/\beta]/(8EI\varepsilon_2^2\alpha) \quad ,$$

where the upper sign is appropriate to the case $z > 0$ and the lower to $z < 0$.

Substituting $z = 0$ into the above expression we get the displacement under the force

$$w(z = 0) = F/(8EI\varepsilon_2^2\alpha) \quad .$$

With the help of (4.184) and (4.181) this may be rewritten as

$$w(z=0) = F(4\chi^3 EI)^{-1/4}/(1 - v^2/c_{\min}^2)^{1/2} \quad , \tag{4.186}$$

which shows a resonance when the velocity of the load approaches the minimal phase velocity of free waves. The value $v = c_{\min}$ is therefore referred to as the critical velocity. This problem, along with that of the previous section, displays a variety of ways the intrinsic features of waveguides may influence their dynamic response.

Problems

4-1 Show that the boundary conditions cannot, in general, be satisfied if either P- or S-reflected waves are ignored.

4-2 Analyze the reflection of SH- and SV-waves at a free surface.

4-3 Why do the equations of motion not appear in derivation of (4.1) – (4.5)? Is the shear wave considered in Section 4.1 of SV- or SH-mode?

4-4 Find the energy partition between the reflected P- and SV-waves making use of the results of Section 4.1.

4-5 Propose a procedure for analysis of grazing incidence of P-waves and derive the relevant results.

4-6 Consider the reflection of P-waves at a rigid surface.

4-7 Analyze the total reflection of SV-waves at a free surface.

4-8 Derive the basic results for reflection of viscoelastic waves at an interface.

4-9 The split Hopkinson pressure bar consists of a short cylindrical specimen placed between two long rods made of high-strength steel (Figure 4-24). When the incident wave strikes the junction some of its energy is reflected from the specimen and some is transmitted to the second rod. Deduce the stress-strain relation while neglecting the wave propagation in the specimen.

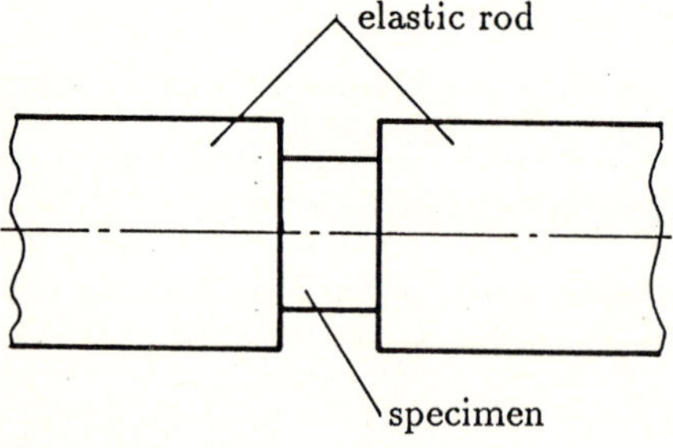

Fig. 4-24 Hopkinson pressure bar.

4-10 Propose a theory of random Rayleigh waves in viscoelastic half-space.

4-11 Outline a way of extending the analysis of Section 4.4 to a piezoelectric medium.

4-12 The layered structure depicted in Figure 4-9 is in contact with an ideal fluid. Will it affect the Love waves propagation?

4-13 Find the impulse response of the interdigital transducer considered in Section 4.6.

4-14 Compute the phase velocity associated with (4.87) for $\gamma = k = 0$ and interpret the results.

4-15 Analyse SH-waves propagation in an infinite plate.

4-16 Derive the group velocity, c_g, for flexural waves in rods governed by the elementary theory and compare it with the exact solution.

4-17 Derive radiation from an oscillating dislocation which is placed eccentrically in a circular cylinder.

4-18 Derive the solution to the problem of the moving load (Section 4.17) for the case $v > \varepsilon_2/\varepsilon_1$.

References and Additional Reading

Sections 4.1 – 4.2

Achenbach (1973), Kolsky (1967), Miklowitz (1978).

Sections 4.3 – 4.6

Dieulesaint and Royer (1980) and Viktorov (1967) present an extensive coverage of surface waves. See Beltzer (1981) for a theory of random Rayleigh waves.

Sections 4.7 – 4.13

Achenbach (1973), Graff (1975), Meeker and Meitzler (1964), Miklowitz (1978), Mindlin (1951).

Sections 4.14 – 4.15

The presentation follows Reismann (1967). See also Hutchinson (1984), Pursey (1957), and Reismann (1968) for related problems.

Section 4.16

Beltzer (1983a). See Beltzer and Parnes (1984) and Parnes and Beltzer (1984, 1986) for a solution which takes into account the eccentricity in the dislocation location and viscoelastic losses.

Section 4.17

Biderman (1972), Fryba (1973).

Chapter V

Wave-Obstacle Interactions. Waves in Composites

The notion of homogeneity is in fact a convenient idealization, which may not be always admissible. The well-known fact is that inhomogeneity of composites has a profound influence on their response. Even such materials as metals are heterogeneous at sufficiently small scale.

We first consider the phenomena induced by the wave interaction with an obstacle, such as diffraction, refraction, scattering, and stress concentrations. This, besides the interest of its own, paves a way for a better understanding of the mechanics of composites.

Then investigations of the static response of composites are given, which are often sufficient for evaluation of low-frequency waves propagation. We present some of the approximate techniques, as well as exact definitions and rigorous bounds. Then an essentially dynamic theory is given, which provides the response of random composites for the entire frequency interval, $0 \leq \omega < \infty$. Considerations of basic features of wave propagation in periodic systems complete the chapter. It should be noted that investigations of ordered structures may be carried out in the spirit of classic methods of elasticity and may be crowned with the exact results. On the other hand, investigations of random composites, which are essentially approximate, require particular attention to their proper interpretation. In fact, these solutions are influenced by the amount and the nature of information concerning phase geometry.

Unfortunately, problems dealing with wave interaction with a crack have not been treated due to the specific nature of this subject. Its presentations would require a change in the basic approach, which is mainly concerned with the overall, not local phenomena. Nevertheless, the effective response of cracked solids is amenable to solution by the presented methods.

5.1 Wave-Obstacle Interactions

In homogeneous, isotropic, infinitely extended media an elastic disturbance propagates without any distortion along a path due to the complete equivalency of the points and the directions. This situation changes, if the disturbance meets an obstacle, say, an insertion with properties different from those of the host medium.

The wave-obstacle interaction gives rise to three main phenomena of interest. First, the wave deviates from its original path, which is referred to as diffraction. Second, excited by the incident wave, the obstacle becomes a secondary source of radiation, or of the so-called scattered waves running in the surrounding medium, and the refracted waves, running in the obstacle. And, finally, since the total field now consists of two parts, the incident and scattered disturbances, the stresses may differ from those in the homogeneous medium, in vicinities of the obstacle they may take on substantially higher values. This effect is known as dynamic stress concentrations.

To illustrate these phenomena consider SH-waves impacting a long cylindrical inclusion in an elastic medium. Accordingly, use can be made of the results of Section 2.5.

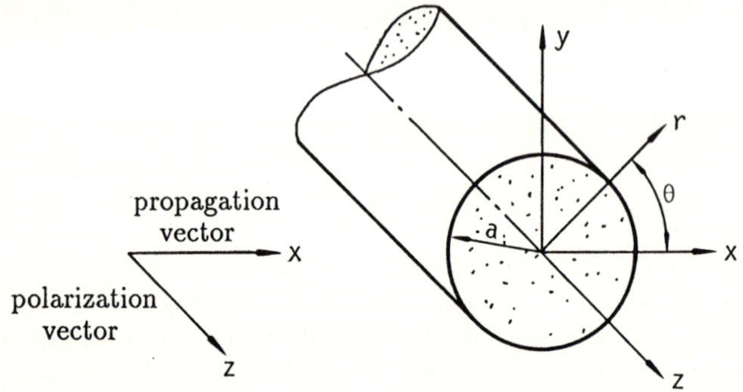

Fig. 5-1 SH-waves interaction with a cylindrical inclusion.

Figure 5-1 shows the system at hand. The cylinder, with radius a, the shear modulus μ_2, and the density ρ_2, is embedded in an elastic medium with the shear modulus μ_1, and the density ρ_1. The incident wave of shear

$$\begin{aligned} u_z^i &= u_o^i e^{i(k_{\beta 1}x-\omega t)} \\ u_x^i &= u_y^i = 0, \qquad k_{\beta 1} = \omega / c_{\beta 1} \end{aligned} \tag{5.1}$$

propagates in the medium and impacts the inclusion. Such a wave may be excited, for example, by oscillating tractions distributed over a vertical plane far away from the inclusion.

This wave gives rise to the scattered waves (propagating in the medium) and the refracted waves (propagating in the inclusion). It is natural to assume that these are also of sole shear,

$$u_z^s = u^s(x,y)e^{-i\omega t}, \qquad u_z^r = u^r(x,y)e^{-i\omega t} \quad , \tag{5.2}$$

where superscripts s and r denote the scattered and refracted fields, respectively.

The explicit expressions for u_z^s and u_z^r may be found by expanding the displacements, as given in Section 2.5, and incorporating the boundary conditions at the interface. To this end, u_z^i is represented, using (5.1) and (2.58), by

$$u_z^i = u_o^i \sum_{n=0}^{\infty} \xi_n i^n J_n(k_{\beta 1} r) \cos n\theta e^{-i\omega t} \quad , \tag{5.3}$$

where the polar coordinates shown in Figure 5-1 have been employed and

$$\xi_n = \begin{cases} 1, & n = 0 \\ 2, & n \geq 1 \end{cases} \quad .$$

In order to represent the scattered field, the outgoing waves only should be incorporated. Hence, on adopting (2.52), we get

$$u_z^s = u_o^i \sum_{n=0}^{\infty} A_n H_n^{(1)}(k_{\beta 1} r) \cos n\theta e^{-i\omega t} \quad , \tag{5.4}$$

where A_n are to be found from the boundary conditions. On the other hand, the refracted waves are shear disturbances running back and forth inside the inclusions. Thus, a solution of standing waves, given by (2.55), is appropriate. However, the Bessel functions of the second kind, $Y_n(q)$, have a singularity when $q \to 0$ in accordance with small-argument behavior

$$\begin{aligned} &\lim_{q \to 0} Y_o(q) = 2 \ln q / \pi \\ &\lim_{q \to 0} Y_n(q) = -(2/q)^n (n-1)!/\pi, \quad \text{for } n > 1 \quad . \end{aligned} \tag{5.5}$$

Therefore, the presence of these functions has no physical basis in the case at hand and the coefficients B_n in (2.55) must be set to zero. Since no dependence on z is anticipated, $\gamma_1 = 0$, and (2.55) takes eventually the form

$$u_z^r = u_o^i \sum_{n=0}^{\infty} C_n J_n(k_{\beta 2} r) \cos n\theta e^{-i\omega t} \quad , \tag{5.6}$$

with $k_{\beta 2} = \omega / c_{\beta 2}$, $c_{\beta 2}^2 = \mu_2 / \rho_2$, and C_n unknown coefficients.

It remains to identify the coefficients A_n and C_n appearing in (5.4) and (5.6) from the boundary conditions at $r = a$. The relation

$$u_z^i(a, \theta) + u_z^s(a, \theta) = u_z^r(a, \theta) \tag{5.7}$$

matches the displacement in the matrix with that in the inclusion at the common boundary. Similarly,

$$\sigma_{rz}(a, \theta) = \mu_1 \left[u_{z,r}^i(a, \theta) + u_{z,r}^s(a, \theta) \right] = \mu_2 u_{z,r}^r(a, \theta) \tag{5.8}$$

implies equal stress across the boundary. Other stresses, associated with this surface, vanish.

Substitutions of (5.3), (5.4), and (5.6) in (5.7) and (5.8) yield the unknown coefficients

$$\begin{aligned} A_n &= (-i^n \xi_n / \delta) \left[\mu_2 k_{\beta 2} J_n(q_1) J_n'(q_2) - \mu_1 k_{\beta 1} J_n'(q_1) J_n(q_2) \right] \\ C_n &= (i^n \xi_n / \delta) \mu_1 k_{\beta 1} \left[J\prime_n(q_1) H_n(q_1) - H_n'(q_1) J_n(q_1) \right] \quad , \end{aligned} \tag{5.9}$$

where

$$\delta = \mu_2 k_{\beta 2} H_n(q_1) J_n'(q_2) - \mu_1 k_{\beta 1} H_n'(q_1) J_n(q_2)$$

$$q_1 = k_{\beta 1} a, \quad q_2 = k_{\beta 2} a, \quad Z'(q) = dZ(q)/dq \quad .$$

The solution obtained will be of use in some of the following sections, which deal with stress perturbations and scattered energy due to the wave-obstacle interaction. In particular, the effect of stress concentrations and the concept of scattering cross-sections will be considered in some detail.

5.2 Dynamic Stress Concentrations

One of the consequences of wave-obstacle interactions relevant for the strength of solids is dynamic stress concentration. The response of a cylindrical cavity under the excitation by SH-waves provides a simple illustration. It can be recovered from results of the previous section by stipulating $\mu_2 = \rho_2 = 0$.

Then the stresses are given by

$$\sigma_{zr} = \mu_1 u_o^i / r \sum_{n=0}^{\infty} \Big\{ \xi_n i^n [n J_n(k_{\beta 1} r) - k_{\beta 1} r J_{n+1}(k_{\beta 1} r)] + $$

$$+ A_n [n H_n(k_{\beta 1} r) - k_{\beta 1} r H_{n+1}(k_{\beta 1} r)] \Big\} \cos n\theta e^{-i\omega t} \tag{5.10}$$

$$\sigma_{\theta z} = -\mu_1 u_o^i / r \sum_{n=0}^{\infty} n \Big[\xi_n i^n J_n(k_{\beta 1} r) + A_n H_n(k_{\beta 1} r) \Big]$$

$$\sin n\theta e^{-i\omega t} \quad ,$$

with

$$A_n = -\xi_n i^n \frac{J_n'(q)}{H_n'(q)} = -\xi_n i^n \frac{n J_n(q) - q J_{(n+1)}(q)}{n H_n(q) - q H_{(n+1)}(q)} \quad ,$$

where $q = k_{\beta 1} a$ and $H(q)$ is the Hankel function of the first kind. At the surface of the cavity, $r = a$,

$$\sigma_{rz} = 0$$

$$\sigma_{\theta z} = -\sigma_o^i e^{-i(\omega t - \pi/2)} \sum_{n=0}^{\infty} (\xi^n i^{n-1} n/q)$$

$$[J_n(q) - J_n'(q) H_n(q)/H_n'(q)] \sin n\theta \quad , \tag{5.11}$$

with $\sigma_o^i = k_{\beta 1} \mu_1 u_o^i$.

On the other hand, in the absence of a cavity, the stress is given everywhere by the incident wave solely. The only non-zero stress is

$$\sigma_{xz}^i = \mu_1 u_{z,x}^i = \mu_1 k_{\beta 1} u_o^i e^{i(k_{\beta 1} x - \omega t)} = \sigma_o^i e^{i(k_{\beta 1} x - \omega t + \pi/2)} \tag{5.12}$$

or in polar coordinates,

$$\sigma^i_{rz} = \sigma^i_o \cos\theta e^{i(k_{\beta 1}x - \omega t + \pi/2)}$$
$$\sigma^i_{\theta z} = -\sigma^i_o \sin\theta e^{i(k_{\beta 1}x - \omega t + \pi/2)} \quad . \tag{5.13}$$

Thus, to evaluate the stress concentration effect due to the cavity one should compare (5.11) and (5.13). This is carried out in the remainder of this section.

Either the real or imaginary part of (5.11) represents the actual solution. Choosing, for example, the real part, one gets

$$\sigma_{\theta z} = N_r \cos(\omega t - \pi/2) + N_i \sin(\omega t - \pi/2) \quad ,$$

where N_r and N_i are the real and imaginary parts of the amplitude function in (5.11).

Since $T = 2\pi/\omega$, this relation shows that N_r is the stress at $t = T/4$, whereas N_i is the stress at $t = T/2$. The maximum value is given by the magnitude

$$|\sigma_{\theta z}| = (N_r^2 + N_i^2)^{1/2} \quad .$$

It is natural to define the value

$$|\sigma_{\theta z}/\sigma^i_o| \tag{5.14}$$

as the dynamic stress concentration factor.

Figure 5-2, computed from (5.11), shows this value as a function of θ. The dimensionless wave number, $k_{\beta 1}a = q$, plays the role of a parameter. For frequencies given by $k_{\beta 1}a > 0.1$, the distribution becomes explicitly asymmetric. For $q = 2$ the stress on the illuminated part of the cavity ($\pi/2 < \theta < \pi$) is greater than that on the shadow side ($0 < \theta < \pi/2$).

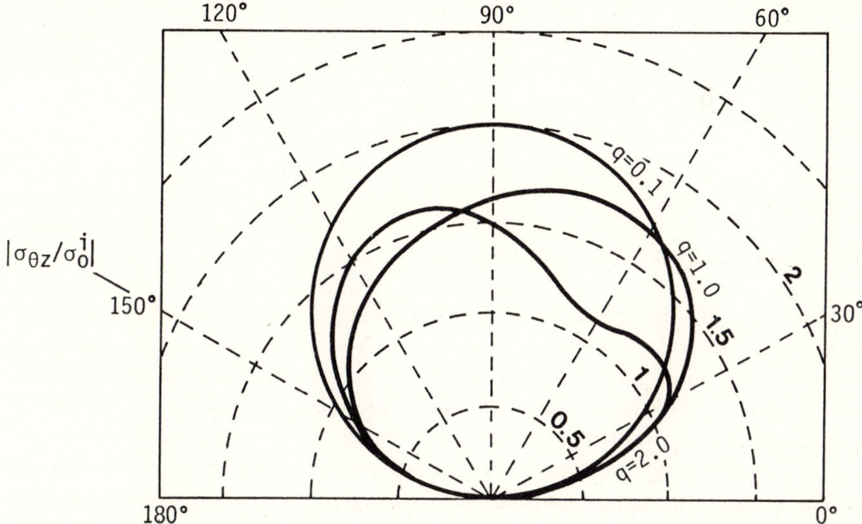

Fig. 5-2 Dynamic stress concentration factor vs. angle for different frequencies.

Figure 5-3 presents the frequency dependence in more detail. It shows $\mathrm{Re}(\sigma_{\theta z}/\sigma_o^i)$ and $\mathrm{Im}(\sigma_{\theta z}/\sigma_o^i)$ versus $k_{\beta 1}a$, for the point $r = a$, $\theta = \pi/2$. The maximum stress occurs at $k_\beta a \simeq 0.4$ and is approximately equal to 2.1.

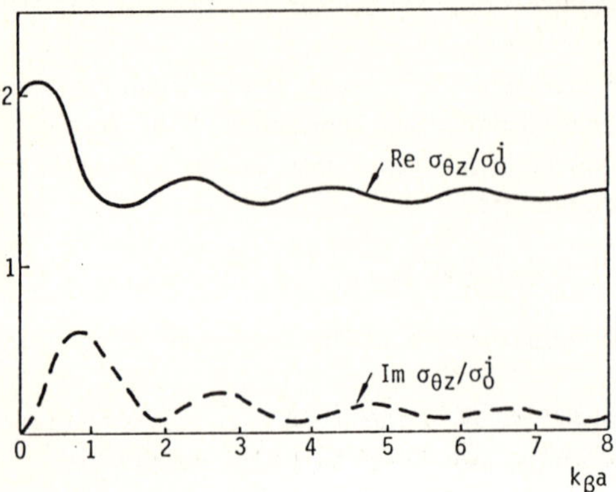

Fig. 5-3 Stress concentration vs. frequency, $\theta = \pi/2$.

Clearly, this analysis furnishes also the solution to the corresponding static problem, which may be recovered from the low-frequency approximations

$$\begin{aligned} J_n(q) &\to q^n/(2^n n!), & n \geq 1 \\ H_n(q) &\to -i2^n(n-1)!/(\pi q^n), & n \geq 1 \end{aligned} \tag{5.15}$$

Thus, when $q \to 0$, the stress at $r = a$ is

$$\sigma_{\theta z}(a, \theta) \simeq 2\sigma_o^i \sin\theta \quad ,$$

which shows that the static stress concentration factor equals 2 for $\theta = \pi/2$.

The strength of the dynamic effect on stress concentrations depends, of course, on a particular system under consideration and may be quite substantial. Along with the resonant waveguides treated in Sections 4.16 and 4.17, it again illustrates a radical influence of dynamic phenomena on the strength of solids.

5.3 Scattering Cross-Sections

An obstacle illuminated by incoming disturbances becomes a source of radiation known as scattered waves. To go on with investigations of basic effects of scattering we must introduce the entities, which would enable us a quantative description. One of the examples of such an approach has been given in the previous section by introducing the factor of dynamic stress concentration. Now we go over to the analysis of the elastic energy perturbations due to wave-obstacle interactions. These have applications in non-destructive testing of materials and provide a basis for appreciating quite surprising effects of wave propagation in random media to be treated later.

5.3.1 Definitions

Our analysis will make use of some of the results of Section 2.13, which are recalled below. The elastic energy transfer rate per unit area in the direction of propagation, m_i, or the so-called intensity, e, is

$$e = P_i m_i \quad , \tag{5.16}$$

where P_i is the Poynting vector. Similarly, the power flux over a closed surface S is given by

$$I = \int_S \int P_i n_i ds \tag{5.17}$$

and the average power flux by

$$\langle I \rangle = \frac{1}{T} \int_0^T I dt \quad , \tag{5.18}$$

in accord with (2.133). Here T is a typical time interval, which for harmonic waves is equal to $2\pi/\omega$ and n_i is normal to ds.

Now we consider an obstacle illuminated by incoming plane harmonic waves. The wave-obstacle interaction gives rise to the scattered field with the average power flux,

$$\langle I^s \rangle = \frac{1}{T} \int_0^T \int_S \int P_i^s n_i ds dt \quad , \tag{5.19}$$

as follows from (5.18). This quantity describes the energy scattered into all directions and, thus, lost by an incident beam at the expense of its interaction with the inclusion. The above surface, S, should be associated with the surface of the obstacle. Nevertheless, no power loss occurs in the elastic medium, and the integral may be taken over another more suitable surface. It might be particularly convenient to apply the far-field approximation, $r \to \infty$.

The magnitude of $\langle I^s \rangle$ depends, among others, upon the average incident intensity, $\langle e^i \rangle$. We therefore introduce the quantity

$$\gamma(\omega) = \langle I^s \rangle / \langle e^i \rangle \tag{5.20}$$

to characterize the sole effect of the obstacle on the power withdrawn from the incident elastic wave. The value, $\gamma(\omega)$, is referred to as the total elastic scattering cross-section.

As an integral entity, $\gamma(\omega)$ may turn out too "crude" as a quantitative measure of scattering, in particular, it does not describe the angular distribution of the scattered power. To this end, we obtain with the help of (5.19) the "portion" scattered across the surface element, ds,

$$d\gamma = \langle n_i \, P_i^s \rangle ds / \langle e^i \rangle \quad ,$$

where n_i is normal to ds. Since

$$ds = r^2 d\Omega \quad ,$$

where $d\Omega$ is the differential element of solid angle and

$$P_i^s = -\sigma_{ij}^s \dot{u}_j^s$$

we get

$$d\gamma/d\Omega = -\langle n_i \sigma_{ij}^s \dot{u}_j^s \rangle r^2 / \langle e^i \rangle \quad .$$

In the three-dimensional case the fields, σ_{ij}^s and $\dot{u}_j^s$, are of the order $1/r$ as $r \to \infty$. This enables us to characterize the angular distribution by the following value

$$\Gamma(\omega, \Omega) = \lim_{r\to\infty} \frac{d\gamma}{d\Omega} = -\lim_{r\to\infty} \frac{\langle n_i \sigma_{ij}^s \dot{u}_j^s \rangle r^2}{\langle e^i \rangle} \quad , \tag{5.21}$$

which is called the differential elastic cross-section.

The entities, $\gamma(\omega)$ and $\Gamma(\omega, \Omega)$, are useful for identification of the volume and the shape of inclusion by measurements made far away from it. This capability is utilized in the non-destructive testing of materials.

The analysis of power loss due to elastic scattering by an obstacle extends to more complicated cases. If the obstacle is made of a viscoelastic material, then, in addition to scattering losses, there is absorption inside the scatterer. Another case, which is of particular interest for solid media, is a viscoelastic matrix containing a purely elastic inclusion. A possible modification of (5.20) is given in the comments on Problem 5-4.

Scattering is substantially governed by the ratio of the wavelength, λ, to a typical size of the obstacle, a. It is called the Rayleigh scattering, if $\lambda/a \gg 1$, and geometric, or Mie scattering, if $\lambda/a \ll 1$. Usually, but not always,

$$\gamma(\omega) \sim 0(\omega^4), \qquad \text{as} \qquad \lambda/a \gg 1 \tag{5.22}$$

in the three-dimensional case, and

$$\gamma(\omega) \sim 0(\omega^3), \qquad \text{as} \qquad \lambda/a \gg 1 \tag{5.23}$$

in the two-dimensional case. Next, denote the cross-section of an obstacle, normal to the incident beam, as A. Then

$$\gamma(\omega) \sim 2A, \qquad \text{as} \qquad \lambda/a \ll 1 \quad , \tag{5.24}$$

The above laws were found in experimental and theoretical ways for many cases of acoustic, electromagnetic, and elastic waves, although the general formal proof is not always available. Equation (5.24) follows also from the picture of geometric optics, if we take into consideration that the scattered field occurs not only in the illuminated, but in the shadow zone too, where it must be intensive enough to cancel out the incident field. These results are exemplified by the following problem.

5.3.2 Scattering Cross-Section of a Cavity

Consider the computation of $\gamma(\omega)$ for the case of a cavity treated in Section 5.2. The average intensity of the incident wave follows from averaging (5.16)

$$\langle e^i \rangle = -(1/4)i\omega(\sigma^o_{xz}u^{o*}_z - \sigma^{o*}_{xz}u^o_z) \quad ,$$

where, according to (5.12) and (5.1),

$$\sigma^o_{xz} = \sigma^i_o e^{i(k_{\beta 1}x+\pi/2)}$$

$$u^o_z = u^i_o e^{ik_{\beta 1}x}$$

and asterisk denotes complex conjugate. Hence,

$$\langle e^i \rangle = (1/2)\mu_1 k_{\beta 1}\omega(u^i_o)^2 \quad , \tag{5.25}$$

which is similar to (2.137), written for dilatational waves.

Turning to the scattered waves, choose a cylindrical surface, $r = R$. Then the average power flux of the scattered field is

$$\langle I^s \rangle = -(1/4)i\omega \int_0^{2\pi} (\sigma^o_{rz}u^{o*}_z - \sigma^{o*}_{rz}u^o_z)^s \Big|_{r=R} R d\theta \quad , \tag{5.26}$$

where (2.134) has been used. Here σ^o_{rz} and u^o_z are the amplitudes associated solely with the scattered field. This is denoted by superscript, s, placed by the bracketed term. These amplitudes are given, according to (5.4) and (5.10), by

$$\sigma^o_{rz} = \mu_1 u^i_o/r \sum_{n=0}^{\infty} A_n[nH_n(k_{\beta 1}r) - k_{\beta 1}rH_{n+1}(k_{\beta 1}r)]\cos n\theta$$

$$u^o_z = u^i_o \sum_{n=0}^{\infty} A_n H_n(k_{\beta 1}r)\cos n\theta \quad .$$

Since

$$\int_0^{2\pi} \cos n\theta \cos m\theta d\theta = \pi\delta_{mn} \qquad \text{for} \qquad m \neq n \neq 0 \quad ,$$

substitution of the above relations into (5.26) yields

$$\langle I^s \rangle = -(1/4)i\pi\mu_1\omega(u^i_o)^2 \sum_{n=0}^{\infty}(2/\xi_n)|A_n|^2 k_{\beta 1}R$$
$$\Big[H'_n(k_{\beta 1}R)H^*_n(k_{\beta 1}R) - H'^*_n(k_{\beta 1}R)H_n(k_{\beta 1}R)\Big] \quad .$$

Next, invoking the relation

$$H^{(1)*}_n(q) = H^{(2)}_n(q)$$

and the Wronksian

$$W\Big[H^{(1)}_n(q),\ H^{(2)}_n(q)\Big] = -4i/(\pi q) \quad ,$$

we eventually get

$$\langle I^s \rangle = \mu_1 k_{\beta 1} c_\beta (u_o^i)^2 (2|A_o|^2 + \sum_{n=0}^{\infty} |A_n|^2) \quad .$$

Note that the far-field approximation for (5.26) would provide the same result. It remains to divide this relation by (5.25) to obtain

$$\gamma(\omega) = (2/k_{\beta 1})(2|A_o|^2 + \sum_{n=1}^{\infty} |A_n|^2) \quad , \tag{5.27}$$

with A_n given in the previous section.

For low-frequencies, $k_{\beta 1} a \ll 1$, the above relation can be simplified by making use of small-argument expansions of the Bessel (Hankel) functions. This yields

$$A_o \simeq -i\pi (k_{\beta 1} a)^2/4$$

$$A_n \simeq i^{n+1} 2\pi (k_{\beta 1} a/2)^{2n}/[n!(n-1)!], \quad n \geq 1 \quad .$$

After substitution of these expressions in (5.27) the scattering cross-section in the Rayleigh approximation is found to be

$$\gamma(\omega) \simeq 3\pi^2 (k_{\beta 1} a)^4 c_{\beta 1}/(4\omega) = 3\pi^2 \omega^3 a^4/(4c_{\beta 1}^3) \quad , \tag{5.28}$$

in accord with (5.23).

The expansion for the case of high-frequencies (Mie scattering) is far more involved and will not be given here. It provides

$$\lim_{\omega \to \infty} \gamma(\omega) = 4a \quad ,$$

which is in agreement with (5.24).

5.4 Diffraction by a Sphere

The diffraction of harmonic P-waves by a spherical inclusion of radius a embedded in an elastic medium is a classic problem of wave-obstacle interaction (see Figure 5-4). In what follows index 1 will again denote matrix and 2 inclusion material.

Let the incident wave

$$u_z^i = u_o^i e^{i(k_{\alpha 1} z - \omega t)} \tag{5.29}$$

propagate along the z-axis and illuminate the inclusion surface, $r = a$.

In view of separability of the wave equation in spherical coordinates, one expects the normal mode approach given in Section 2.6 to furnish the solution. Preliminary analysis shows that the field has no dependence on the angle ϕ because of symmetry. Further, by the same argument, the displacement u_ϕ must be set to zero. The wave field is characterized therefore by the functions u_r and

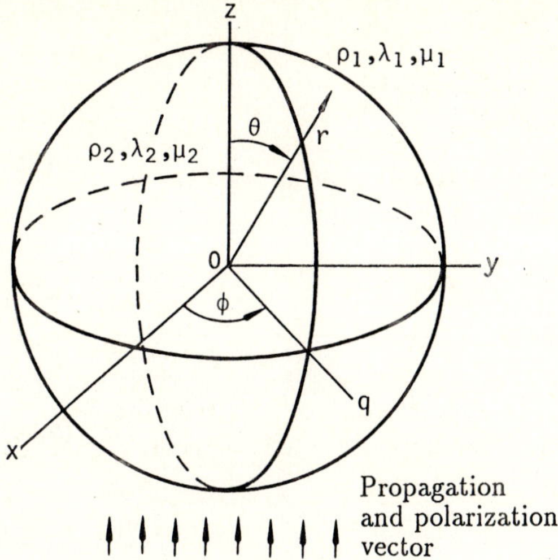

Fig. 5-4 Diffraction by a spherical obstacle.

u_θ, which depend on spatial coordinates r and θ. Equation (2.70) indicates that the wave potentials are

$$\Pi = \Pi(r,\theta), \quad \chi = \chi(r,\theta), \quad \psi = 0 \quad .$$

Here and henceforth we may often omit the multiplier $e^{-i\omega t}$.

Since $z = r\cos\theta$, the incident wave may be rewritten as

$$u_z^i = u_o^i e^{i(k_{\alpha 1} r\cos\theta - \omega t)} \quad .$$

Recalling (2.68) and (2.69), we get

$$u_z^i = u_o^i \sum_{n=0}^{\infty} (2n+1) i^n j_n(k_{\alpha 1} r) P_n(\mu) \tag{5.30}$$

or in terms of the incident wave potential

$$\Pi^i = \Pi_o^i \sum_{n=0}^{\infty} (2n+1) i^n j_n(k_{\alpha 1} r) P_n(\mu) \quad , \tag{5.31}$$

where, as (2.39) shows,

$$\Pi_o^i = u_o^i/(ik_{\alpha 1}) \tag{5.32}$$

and $\mu = \cos\theta$.

Turning to the potentials of the scattered and refracted waves, we apply (2.66), which represents an elementary spherical wave symmetric with respect to the z-axis. Hence,

$$\begin{aligned} \Pi^s &= \sum_{n=0}^{\infty} A_n h_n^{(1)}(k_{\alpha 1} r) P_n(\mu) \\ \chi^s &= \sum_{n=0}^{\infty} B_n h_n^{(1)}(k_{\beta 1} r) P_n(\mu) \end{aligned} \tag{5.33}$$

and

$$\begin{aligned} \Pi^r &= \sum_{n=0}^{\infty} C_n j_n(k_{\alpha 2} r) P_n(\mu) \\ \chi^r &= \sum_{n=0}^{\infty} D_n j_n(k_{\beta 2} r) P_n(\mu) \quad , \end{aligned} \tag{5.34}$$

where superscripts s and r denote the scattered and refracted waves, and A_n, B_n, C_n, and D_n are to be specified from the boundary conditions. In writing (5.33), we have chosen the spherical Hankel functions of the first kind, $h_n^{(1)}(q)$, to represent outgoing disturbances emanating from the inclusion. The refracted waves, confined in the inclusion, are taken as standing waves. Note that the same reasoning has been applied to deduce (5.4) and (5.6), which deal with a cylindrical scatterer.

For an elastic sphere, firmly bonded with the matrix, the boundary conditions match the displacements and stresses across the interface, $r = a$,

$$\begin{aligned} u_r^i + u_r^s = u_r^r, \quad \sigma_{rr}^i + \sigma_{rr}^s = \sigma_{rr}^r, \qquad \text{at} \qquad r = a \\ u_\theta^i + u_\theta^s = u_\theta^r, \quad \sigma_{r\theta}^i + \sigma_{r\theta}^s = \sigma_{r\theta}^r, \qquad \text{at} \qquad r = a \quad . \end{aligned} \tag{5.35}$$

Four equations (5.35) enable us to find the four coefficients, A_n, B_n, C_n, and D_n for $n = 0, 1, 2...$, which completes the solution. Although these expressions are tractable, they are too lengthy and standard and need not be given here. Nevertheless, the long-wavelength approximation, which can be derived by a resort to small-argument behavior of Hankel functions, is fairly simple and is given below

$$\begin{aligned} u_r^s &= -iN_1(k_{\alpha 1} a)^3/(k_{\alpha 1}^2 r) \\ &\quad [N_2 - N_3 \cos\theta - N_4(3\cos 2\theta + 1)/4] e^{ik_{\alpha 1} r} \\ u_\theta^s &= -iN_1(k_{\beta 1} a)^3/(k_{\alpha 1} k_{\beta 1} r) \\ &\quad [N_3 \sin\theta + 3k_{\beta 1} N_4 \sin 2\theta/(4k_{\alpha 1})] e^{ik_{\beta 1} r}, \end{aligned} \tag{5.36}$$

with

$$\begin{aligned} N_1 &= ik_{\alpha 1} u_o^i \\ N_2 &= (K_1 - K_2)/(3K_2 + 4\mu_1) \\ N_3 &= (\rho_1 - \rho_2)/(3\rho_1) \end{aligned} \tag{5.37}$$

$$N_4 = \frac{20}{3}\frac{\mu_1(\mu_2-\mu_1)}{6\mu_2(K_1+2\mu_1)+\mu_1(9K_1+8\mu_1)} \quad .$$

Here K_1 and K_2 are the bulk moduli of the matrix and inclusion, respectively. The above expressions hold provided the wavelengths of all waves are much longer than the sphere radius, a. The neglected terms are of order $(k_{\alpha 1}a)^5$.

With the help of (5.20), (5.36), and (5.37) we may obtain the total scattering cross-section of an elastic sphere in the Rayleigh region, which is

$$\gamma(\omega) = 4\pi k_{\alpha 1}^4 a^6 g/9 \quad , \tag{5.38}$$

with

$$g = \left[\frac{3\chi_{o1}^2}{(3\chi_{o2}^2-4)M+4}-1\right]^2 + \frac{1}{3}\left[1+2\chi_{o1}^3\right]\left[\left(\frac{c_{\beta 1}}{c_{\beta 2}}\right)^2 M-1\right]^2 + 40(2+3\chi_{o1}^5)\left[\frac{M-1}{2(3\chi_{o1}^2+2)M+(9\chi_{o1}^2-4)}\right]^2 \tag{5.39}$$

and

$$\chi_{o1} = c_{\alpha 1}/c_{\beta 1}, \quad \chi_{o2} = c_{\alpha 2}/c_{\beta 2}, \quad M = \mu_2/\mu_1 \quad .$$

The solution for harmonic waves makes possible investigations of diffraction of transient deterministic or random disturbances. In fact, these are available in literature.

5.5 Response of Random Composites

Having estimated the main effects of wave interaction with a single obstacle, we turn to the response of composite media. These may be divided into two groups: periodic and random. A monocrystal examined on a sufficiently small scale exemplifies the first case, while a polycrystal the second.

Elastic moduli and the density of a heterogeneous medium may spatially vary either in a smooth way or abruptly. The latter case is usually far more important for applications, and will therefore be considered here.

Analysis of periodic composites is quite different from that of random ones, the latter being far more complicated, at least from the logical point of view, than the former. In fact, the properties of periodic multiphase media are given with absolute precision, if we neglect natural scatter of physical and geometrical parameters. Accordingly, an exact solution may be pursued. On the other hand, statistical information of higher order concerning the geometry of random heterogeneous media is difficult to obtain. The analysis and its results are thus influenced by the extent of our preliminary knowledge of the microstructure, which motivates a resort to approximate solutions.

We are thus interested in an overall, average response of random media, which incorporates only the main phenomena taking place, rather than in numerous

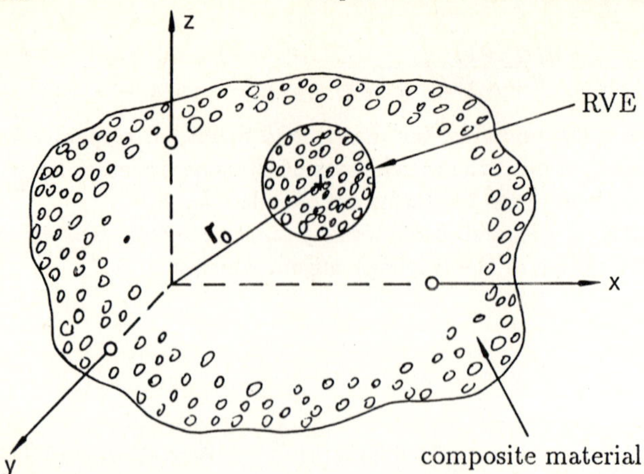

Fig. 5-5 The representative volume element (RVE).

details. One of the possible approaches is to construct an effective homogeneous medium, which effectively responds like the original inhomogeneous one.

To make it clearer, consider, as an example, material consisting of spherical inclusions randomly dispersed in a homogeneous matrix. For such materials it is possible to define the so-called representative volume element (RVE), as shown in Figure 5-5, which is large compared to typical phase region dimensions but is still small compared to the entire body. The material may be referred to as statistically homogeneous provided that statistical moments of the fields (see Section 1.3) are the same regardless of the RVE position within the solid. For such materials a physically meaningful identification of their effective response may be developed by introducing the concept of effective homogeneous media. These arise from averaging, in one way or another, the composite properties over the RVE.

Once the effective elastic moduli and the density are known, they may be used for analyzing the behavior of large-scale structures, which are thought of as made of an effective material. It should be stressed that this approach holds for predicting the overall response only, and is inapplicable for investigations of local effects, such as stress concentrations or interface phenomena.

Our previous analysis of wave-obstacle interactions shows that the wave propagation in a composite may depend on the ratio of the wavelength, λ, to a typical dimension of the microstructure, a. In other words, the composite appears as a dispersive medium, even if its constituents are perfectly elastic.

Furthermore, at the expense of scattering by randomly located inclusions part of the energy of the incident wave transfers to incoherent motions. It follows, that a perfectly elastic random composite may be seen by a wave as an attenuative medium, despite the conservative nature of the entire system. The strength of this effect should also depend on the ratio, λ/a, as considerations of the total scattering cross section, $\gamma(\omega)$, show.

It has been noted in the beginning of this section that the incompleteness of information available on the random phase geometry puts a limit on "resolution", which may be achieved. Furthermore, it appears that the theoretical modelling of random composites depends substantially on the nature and amount of this information. This differs from classical problems, like those considered in the previous chapters, which are posed so as to insure the existence of the exact and unique solution.

For the absence of exact evaluations, a particular attention should be paid to correlations among approximate solutions and their consistence with a priori given information or bounds. We shall present therefore several alternative approaches. We begin with modelling the static or long-wavelength response, $\lambda/a \to 0$, which, as the first approximation, is of primary interest.

5.6 Effective Scatterer Approach

Consider a material consisting of identical spherical inclusions with radius, a, which are randomly dispersed in a homogeneous matrix (Figure 5-5). Denote the bulk and shear moduli of the matrix as K_1 and μ_1, and those of inclusions as K_2 and μ_2. The densities are ρ_1 and ρ_2, respectively. It is supposed that the volume fraction, ϕ, is the only information available, which concerns the phase distribution.

Assume that a representative sphere, shown in Figure 5-5, contains N inclusions and is irradiated by incoming wave, $\boldsymbol{u}^i$, given by (5.29). Each of the spherical inclusions inside the representative sphere will generate a scattered field, $\boldsymbol{u}_j^s$, which, in turn, will interact with the fields of other inclusions in a complicated way. If the composite is dilute enough (small values of the volume fraction, ϕ) the multiple interactions effects may be neglected. Then the total field outside the representative sphere is

$$\boldsymbol{u}(\boldsymbol{r}) = \boldsymbol{u}^i(\boldsymbol{r}) + \sum_{j=1}^{N} \boldsymbol{u}_j^s(\boldsymbol{r}, \boldsymbol{r}_j) \quad . \tag{5.40}$$

Here $\boldsymbol{r}$ denotes an observation site and $\boldsymbol{r}_j$ the location of the j-th inclusion.

Looking for a reasonable averaging procedure over the volume of the representative sphere, V_o, we now think of it as it were made of an effective homogeneous material with the constants K_e, μ_e, and ρ_e. Then Figure 5-5 shows a single sphere made of the above material and embedded in the matrix. Accordingly, the total field outside this sphere is

$$\boldsymbol{u}(\boldsymbol{r}) = \boldsymbol{u}^i(\boldsymbol{r}) + \boldsymbol{u}_e^s(\boldsymbol{r}, \boldsymbol{r}_o) \quad , \tag{5.41}$$

where $\boldsymbol{r}_o$ defines the center of representative sphere and $\boldsymbol{u}_e^s$ is the scattered field.

On equating (5.40) and (5.41) we may find the effective parameters, K_e, μ_e, and ρ_e. We have, however, to take care of avoiding the dependence on $\boldsymbol{r}_j$, since statistical information concerning the inclusion location is not available, according to the statement of the problem. To this end, we assume that the observation point, $\boldsymbol{r}$, is far away from $\boldsymbol{r}_o$, so that in the first approximation

$$\boldsymbol{r}_j \simeq \boldsymbol{r}_o \quad . \tag{5.42}$$

We also assume that the wavelengths of all waves are sufficiently small compared to the radius of the representative sphere, to make possible the use of the long-wavelength approximations given by (5.36) and (5.37).

Now we proceed with calculations. The waves scattered by the representative sphere are given by (5.36) and (5.37) with the following replacements made

$$K_2 \to K_e, \quad \mu_2 \to \mu_e, \quad \rho_2 \to \rho_e \tag{5.43}$$

and

$$a \to R = [V_o/(4\pi/3)]^{1/3} \quad ,$$

with R representing radius of this imaginary effective scatterer.

The superposition of the waves scattered by N actual inclusions is

$$\begin{aligned}\sum_{j=1}^{N} u_r^j &= -iN_1 N(k_{\alpha 1}a)^3/(k_{\alpha 1}^2 r)[N_2 - N_3\cos\theta \\ &\qquad -N_4(3\cos 2\theta + 1)/4]e^{ik_{\alpha 1}r} \\ \sum_{j=1}^{N} u_\theta^j &= -iN_1 N(k_{\beta 1}a)^3/(k_{\alpha 1}k_{\beta 1}r)[N_3\sin\theta \\ &\qquad + 3k_{\beta 1}N_4\sin 2\theta/(4k_{\alpha 1})]e^{ik_{\beta 1}r} \quad ,\end{aligned} \tag{5.44}$$

where use has been made of (5.36). The coefficients, N_1, N_2, and N_3, appearing in (5.44) are given by (5.37).

On substituting (5.29), (5.44) and (5.36) (with the appropriate replacements given by (5.43)) into (5.40) and (5.41) and equating them, we note that this equality must hold independently of the angle θ. This implies that the coefficients of the corresponding angular terms are equal and provides thereby the following relations for the effective constants:

$$\frac{K_e - K_1}{3K_e + 4\mu_1} = \phi\frac{K_2 - K_1}{3K_2 + 4\mu_1} \tag{5.45}$$

$$\frac{\mu_e - \mu_1}{6\mu_e(K_1 + 2\mu_1) + \mu_1(9K_1 + 8\mu_1)} = \frac{\phi(\mu_2 - \mu_1)}{6\mu_2(K_1 + 2\mu_1) + \mu_1(9K_1 + 8\mu_1)} \tag{5.46}$$

$$\rho_e - \rho_1 = \phi(\rho_2 - \rho_1) \tag{5.47}$$

where the volume fraction, ϕ, is

$$\phi = Na^3/R^3 \quad .$$

Accordingly, the effective wave speeds in the long-wavelength approximation are

$$\begin{aligned} c_\alpha^e &= [(\lambda_e + 2\mu_e)/\rho_e]^{1/2} = [(K_e + 4\mu_e/3)/\rho_e]^{1/2} \\ c_\beta^e &= (\mu_e/\rho_e)^{1/2} \quad .\end{aligned} \tag{5.48}$$

We note that the expression for the effective density, ρ_e, coincides with that of the average density. It is not always the case. When the matrix is inviscid fluid the same approach yields a different result for ρ_e, which may be attributed to the induced mass effect.

An approximate nature of this technique motivates a search for alternative approaches, some of which are considered below.

5.7 Composite Spheres Assemblage

Another approach resorts to a particular geometry of composites, which, while satisfying the concept of statistical homogeneity, would also facilitate the calculation of the effective parameters. As far as the final results make use of the moduli of the constituents and the volume fraction only, they may apply to other geometries too as approximate evaluations.

A devised model deals with a gradation of sizes of spherical inclusions shown in Figure 5-6. A region of the matrix which is taken associated with each inclusion is shown by broken curves. It is chosen so as to provide the same ratio of radii $a/b = \phi$ for any elementary composite sphere. An effective static response of this model, known as composite sphere assemblage, can be obtained by the analysis of a single composite sphere and then its extension to the entire RVE with the help of energy considerations.

Fig. 5-6 Composite spheres assemblage.

Subject the outer surface of the composite sphere, $r = b$, to hydrostatic pressure

$$\sigma_{rr}(r = b) = p \quad .$$

as shown in Figure 5-7. No angular dependence is present because of the symmetry, which makes an investigation easy. The conditions of equilibrium may be shown to yield the single equation

$$u_{r,rr} + 2u_{r,r}/r - 2u_r/r^2 = 0 \quad ,$$

with u_r denoting radial displacement. The solution to this equation is

$$u_{r\delta} = A_\delta r + B_\delta / r^2 \quad ,$$

where $\delta = 2$ for the core and $\delta = 1$ for the external shell. The coefficient B_2 must be set to zero in order to avoid a singularity at $r = 0$. Then the displacements are

$$u_{r2} = A_2 r$$
$$u_{r1} = A_1 r + B_1 / r^2 \quad ,$$

and the relevant stresses are

$$\sigma_{rr2} = (3\lambda_2 + 2\mu_2) A_2$$
$$\sigma_{rr1} = (3\lambda_1 + 2\mu_1) A_1 - 4\mu_1 B_1 / r^3 \quad .$$

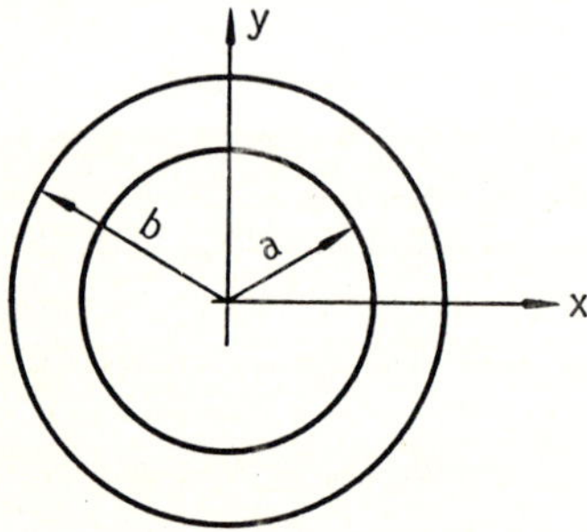

Fig. 5-7 Composite sphere.

Matching the above entities across the interface, $r = a$, we get the constants, A_1, A_2, and B_1 as follows:

$$A_1 = pQ/(4\mu_1)$$
$$B_1 = \left[-p/(4\mu_1) + 3K_1 pQ/(16\mu_1^2) \right] b^3$$
$$A_2 = A_1 + B_1 / a^3 \quad ,$$

where

$$Q = \frac{3K_2 + 4\mu_1}{3(K_2 - K_1)\phi + 9K_2 K_1/(4\mu_1) + 3K_1} \quad .$$

The effective bulk stiffness of the composite sphere, K_e, may be deduced, for example, by comparison of the displacement at the outer surface, $r = b$, with that of the homogeneous sphere subject to the same pressure, p. The latter, u_{rh}, is

$$u_{rh} = pb/(3K) \quad ,$$

which yields

$$K_e = K_1 + \phi(K_2 - K_1)\frac{(4\mu_1 + 3K_1)}{4\mu_1 + 3K_2 + 3(K_1 - K_2)\phi} \quad . \tag{5.49}$$

The same expression follows if one compares the energies stored by the above two spheres.

It remains to extend this result to the entire RVE shown in Figure 5-6. To this end, imagine this volume subjected to the pressure p. The above solution may be viewed as an admissible stress state in the context of the theorem of minimum complementary energy given in Section 1.10. In fact, the above solution applies for each of composite spheres of the RVE by addition of rigid-body translations, which contribute nothing to the strain energy. This provides the lower bound for the stiffness

$$K_{ex} > K_e \quad ,$$

where K_{ex} is the exact stiffness.

Similarly, the problem can be solved for the displacement

$$u_i = Ax_i, \quad A = \text{const}$$

imposed on the external boundary of the composite sphere. This provides, according to the theorem of minimum potential energy, the upper bound for the stiffness. Computations show that this bound coincides with (5.49), which thus turns out to be the exact solution. This is a remarkable feature of the model, which finds applications in interpreting theoretical results to be given later.

It should be stressed that (5.49) is rigorous only for the composite spheres assemblage and must be considered approximate for real materials because of the artificial nature of the underlying model. Furthermore, the exact solution for μ_e for this particular composite, has not been found yet.

Comparison of (5.49) with the previous result, given by (5.45), surprisingly reveals that they are equal. Namely, one of them can be manipulated into another, which supports the adequacy of the approaches. Unlike the effective scatterer approach, the present one is essentially static and may appear simpler. However, it has a more narrow field of applications.

5.8 A Differential Scheme

The main difficulty of the analysis of the effective response is a proper account for mutual interactions among embedded inclusions, which become increasingly important with increase in the volume concentration, ϕ. The above approximate results are therefore less favourable for concentrated mixtures. In fact, (5.45) and (5.46), for example, have been derived while neglecting multiple scattering effects.

Looking for an improved approach, we note that a particular way of producing a composite is not reflected in the governing equations, which, under conventional statements, account for the final phase geometry only. It is therefore possible to devise a construction process, when we start with a matrix, say, 1 and embed inclusions of phase 2 in dilute concentration. This is chosen so as to avoid strong multiple interactions. Next, a "homogenization" is carried by one of the approaches suitable for handling a dilute mixture and then new inclusions are added. This build-up process terminates upon arrival at the final volume fraction, ϕ. This approximate approach is known as a differential scheme, since, in the limit, the increase in the volume fraction, $\Delta\phi$, taking place between two steps, is thought of as a differential

$$\Delta\phi \rightarrow d\phi \quad .$$

Also, the evolution of an effective property of interest is governed by a differential equation.

The effective medium arrived at by the above technique is referred to as a realizable one, in the sense that we can describe an actual construction process of making the composite with the defined behavior. To this end, consider volume V_o of a current "homogenized" matrix obtained after step i and imagine the following removal-replacement procedure for the step $i+1$. Let the volume of the above material be removed by Δv and replaced with the same volume of the inclusion material. The volume of inclusion material after this additional step is

$$V_o\phi_{i+1} = V_o\phi_i - \Delta v\phi_i + \Delta v \quad ,$$

which yields

$$\phi_{i+1} = \phi_i + (\Delta v/V_o)(1-\phi_i) \quad .$$

This may be rewritten in terms of increments

$$\Delta\phi = (1-\phi)\Delta\tilde{\phi} \quad .$$

Here tilda denotes a current value. In the limit $\Delta\phi \rightarrow 0$, $\Delta\tilde{\phi} \rightarrow 0$, we are left with the following relation between the absolute and current increments

$$d\phi = d\tilde{\phi}(1-\phi) \quad , \tag{5.50}$$

with condition $\phi = 0$ at $\tilde{\phi} = 0$.

Given the effective response of a dilute mixture, we may extend the result to more concentrated mixtures by a differential scheme. As an illustration, consider a fiber composite shown in Figure 1-5. The fibers are aligned along the same axis but are randomly dispersed in the transverse plane. The composite thus possesses transverse isotropy and its overall behavior is characterized by five elastic moduli. One of this is responsible for the shear in the x_1x_3-plane (or in the x_2x_3-plane) and is called the axial shear modulus.

To find this modulus we could adopt the concept of effective scatterer and then invoke the analysis of the SH-wave interaction with an embedded elastic cylinder given in Section 5.1 to carry out the calculations. A resort to the idea of composite cylinders, similar to that of composite spheres, is an alternative.

Following the latter approach (see Problem 5-8 and related comments) we get that the effective axial shear modulus, μ_e, is given by

$$\mu_e = \frac{\mu_2(1+\phi)+\mu_1(1-\phi)}{\mu_2(1-\phi)+\mu_1(1+\phi)}\mu_1 \quad . \tag{5.51}$$

Here 1 refers to matrix and 2 to fibres.

In the context of a differential scheme we interpret (5.51) as the "homogenization" operator, applied between steps i and $i-1$. We therefore get

$$\mu_i = \frac{\mu_2(1+\Delta\tilde{\phi})+\mu_{i-1}(1-\Delta\tilde{\phi})}{\mu_2(1-\Delta\tilde{\phi})+\mu_{i-1}(1+\Delta\tilde{\phi})}\mu_{i-1} \quad . \tag{5.52}$$

Setting

$$\mu_{i-1} = \tilde{\mu}, \qquad \mu_i = \tilde{\mu} + \Delta\tilde{\mu}$$

we obtain

$$(\mu_2+\tilde{\mu})\Delta\tilde{\mu} + (\tilde{\mu}-\mu_2)2\tilde{\mu}\Delta\tilde{\phi} = 0 \quad . \tag{5.53}$$

This, in view of (5.50), yields the evolution law for the current effective modulus, $\tilde{\mu}$,

$$\frac{d\tilde{\mu}}{d\phi} = \frac{2\tilde{\mu}(\mu_2-\tilde{\mu})}{(\mu_2+\tilde{\mu})(1-\phi)} \quad . \tag{5.54}$$

Equation (5.54) may be integrated from μ_1 to μ_e, and from 0 to ϕ, respectively, to eventually yield

$$\mu_1(\mu_e-\mu_2)^2 = \mu_e(\mu_1-\mu_2)^2(1-\phi)^2 \quad , \tag{5.55}$$

which defines the effective modulus, μ_e.

Obviously, other results concerning the effective response of composite solid media at dilute concentrations may be improved by a differential scheme. Problem 5-9 provides one of the examples.

Although the conceptual simplicity of the differential scheme is attractive, it is essentially an approximation, which will be further evaluated in Section 5.11. Its dynamic generalization will be given too.

5.9 Response of Polycrystals

Consider, say, a plate made of a polycrystalline material, like metals or alloys. These materials consist of small grains (crystallites), which are chaotically oriented. Since each grain is anisotropic and is seen by a wave under a different angle, the analysis of a plate appears a formidable task, unless we resort to the notion of an effective response in the sense described above.

The evaluation can be carried out by treating the medium as a multiphase one, the inhomogeneity being caused by disordered orientation of crystallites. In this sense, there is an infinite number of phases. Clearly, the symmetry properties of crystallites and a distribution function, which describes their spatial orientation, are the main factors governing the effective response.

5.9.1 Basic Equations

The local behavior is conveniently defined in the reference frame, x_i, which coincides with crystallographic axes of a crystallite, while the overall behavior is defined in a laboratory reference frame, x'_i, suitably placed in a macroscopic specimen. We therefore recall that the above coordinates are related by the rigid-body rotations

$$\begin{aligned} &x'_i = \alpha_{ik} x_k, \qquad x_k = \alpha_{ki} x'_i \\ &\alpha_{in}\alpha_{jn} = \delta_{ij} \quad . \end{aligned} \tag{5.56}$$

Here α_{ik} is the set of direction cosines.

Similarly, for a tensor of the second rank we get

$$T'_{ij} = \alpha_{im}\alpha_{jn}T_{mn}, \qquad T_{ij} = \alpha_{mi}\alpha_{nj}T'_{mn} \quad , \tag{5.57}$$

and for a tensor of the fourth rank

$$T'_{ijk\ell} = \alpha_{ip}\alpha_{jq}\alpha_{kr}\alpha_{\ell s}T_{pqrs}, \quad T_{ijk\ell} = \alpha_{pi}\alpha_{qj}\alpha_{rk}\alpha_{s\ell}T'_{pqrs} \tag{5.58}$$

The above relations enable us to go over from one reference frame to another while dealing, for example, with the tensor of elastic moduli, $c_{ijk\ell}$.

Clearly, a simple intuitive way of evaluating the effective moduli is averaging over all orientation of crystal axes with respect to fixed specimen axes or vice versa, provided the above statistical distribution function is given. For a perfectly disordered polycrystal this function is constant, since all orientations are equally probable, and the effective medium appears as effectively isotropic. The calculations can be carried out with the help of (5.56) – (5.58).

For orthotropic materials the tensor of elastic moduli can be written as

$$\begin{aligned} c_{ijk\ell} = \sum_{n=1}^{3} \Big[&\mu_n(\delta_{in}\delta_{jn}\delta_{k\ell} + \delta_{ij}\delta_{kn}\delta_{\ell n}) + \lambda_n\delta_{in}\delta_{jn}\delta_{kn}\delta_{\ell n} \\ &+ \nu_n(\delta_{in}\delta_{jk}\delta_{\ell n} + \delta_{jn}\delta_{ik}\delta_{\ell n} + \delta_{in}\delta_{j\ell}\delta_{kn} + \delta_{jn}\delta_{i\ell}\delta_{kn})\Big] \quad , \end{aligned} \tag{5.59}$$

where the nine constants, μ_n, ν_n, and λ_n are

$$\begin{aligned} &\mu_1 = (c_{12} + c_{13} - c_{23})/2, \quad \nu_1 = (c_{55} + c_{66} - c_{44})/2, \\ &\mu_2 = (c_{12} + c_{23} - c_{13})/2, \quad \nu_2 = (c_{44} + c_{66} - c_{55})/2, \\ &\mu_3 = (c_{13} + c_{23} - c_{12})/2, \quad \nu_3 = (c_{44} + c_{55} - c_{66})/2, \\ &\lambda_1 = c_{11} + c_{23} + 2c_{44} - (c_{12} + c_{13} + 2c_{55} + 2c_{66}), \\ &\lambda_2 = c_{22} + c_{13} + 2c_{55} - (c_{12} + c_{23} + 2c_{44} + 2c_{66}), \\ &\lambda_3 = c_{33} + c_{12} + 2c_{66} - (c_{13} + c_{23} + 2c_{44} + 2c_{55}) \quad . \end{aligned} \tag{5.60}$$

Since $\alpha_{ki}\delta_{in} = \alpha_{kn}$ and $\alpha_{ki}\alpha_{pi} = \delta_{kp}$, we may readily transform $c_{ijk\ell}$ to the laboratory axes, namely to $c'_{ijk\ell}$, with the help of (5.58) and (5.59),

$$\begin{aligned} c'_{ijk\ell} = \sum_{n=1}^{3} \Big[& \mu_n(\alpha_{in}\alpha_{jn}\delta_{k\ell} + \delta_{ij}\alpha_{kn}\alpha_{\ell n}) + \lambda_n \alpha_{in}\alpha_{jn}\alpha_{kn}\alpha_{\ell n} \\ & + \nu_n(\alpha_{in}\delta_{jk}\alpha_{\ell n} + \alpha_{jn}\delta_{ik}\alpha_{\ell n} + \alpha_{in}\delta_{j\ell}\alpha_{kn} + \alpha_{jn}\delta_{i\ell}\alpha_{kn})\Big] \end{aligned} \tag{5.61}$$

5.9.2 Effective Moduli

Consider an example of a cubic system. Then

$$\begin{aligned} \lambda_1 &= \lambda_2 = \lambda_3 = c_{11} - c_{12} - 2c_{44} \\ \mu_1 &= \mu_2 = \mu_3 = c_{12}/2 \\ \nu_1 &= \nu_1 = \nu_2 = c_{44}/2 \quad , \end{aligned} \tag{5.62}$$

as may be seen from (5.59) and (1.40). The tensor, $c'_{ijk\ell}$, follows from (5.61) and the identity $\delta_{ij} = \alpha_{in}\alpha_{jn}$:

$$\begin{aligned} c'_{ijk\ell} = {} & c_{12}\delta_{ij}\delta_{k\ell} + c_{44}(\delta_{ik}\delta_{j\ell} + \delta_{i\ell}\delta_{jk}) \\ & + (c_{11} - c_{12} - 2c_{44}) \sum_{n=1}^{3} \alpha_{in}\alpha_{jn}\alpha_{kn}\alpha_{\ell n} \quad , \end{aligned}$$

which can be used for the straightforward averaging over the angles of mutual orientation. Some of the useful relations for carrying this out are given in the comments on Problem 5.10.

The straightforward averaging of (5.61) with respect to direction cosines may be quite tedious. If the material is effectively isotropic, then a simpler procedure resorts to general results of the tensor calculus. In fact, for unit tensor of the fourth rank, $\boldsymbol{I}$, we get

$$\boldsymbol{I} = \boldsymbol{V} + \boldsymbol{D} \quad ,$$

where $\boldsymbol{V}$ and $\boldsymbol{D}$ are its "isotropic" (spherical) and "deviatoric" parts given by

$$\begin{aligned} V_{ijk\ell} &= \delta_{ij}\delta_{\ell k}/3 \\ D_{ijk\ell} &= (\delta_{ik}\delta_{j\ell} + \delta_{i\ell}\delta_{jk} - 2\delta_{ij}\delta_{k\ell}/3)/2 \quad . \end{aligned}$$

Then, the tensor of elastic moduli, as another tensor of the fourth rank, can also be decomposed into its isotropic and deviatoric parts as follows:

$$\mathbf{c} = 3K\boldsymbol{V} + 2\mu\boldsymbol{D} \quad , \tag{5.63}$$

where

$$K = c_{iikk}/9, \qquad \mu = (c_{ikik} - c_{iikk}/3)/10 \quad . \tag{5.64}$$

Equation (5.63) resembles a decomposition of Hooke's law for isotropic solids, with K and μ the bulk and shear moduli.

On applying (5.63) and (5.64) to a cubic system and interpreting the results as the effective moduli, we get with the help of (5.59) and (5.62) the following effective moduli

$$K_e = (c_{11} + 2c_{12})/3, \qquad \mu_e = [(c_{11} - c_{12}) + 3c_{44}]/5. \tag{5.65}$$

Other cases of symmetry may be treated in the same way, provided no texture is present.

Angular averaging of the c-tensor is known as the Voight method. By the same argument we may similarly average the tensor of elastic compliances, $g = c^{-1}$, which is referred to as the Reuss method. For a cubic system this yields

$$K_e^{-1} = 3(g_{11} + 2g_{12}), \qquad \mu_e^{-1} = [4(g_{11} - g_{12}) + 3g_{44}]/5 \quad . \tag{5.66}$$

Usually the two ways do not provide the same results, which may be anticipated in view of their approximate nature. Exceptions are the above expressions for K_e, which are identical because $c_{11} + 2c_{12} = 1/(g_{11} + 2g_{12})$ for a cubic symmetry. It can be shown that the two estimates closely relate to bounds on elastic moduli. Further evaluations of the Voight and Reuss estimates will be given in Section 5.11.

A relevant point to note is that the above techniques of estimating the effective static response appeal to a physical intuition rather than to a comprehensive analysis. This appears justified by random microstructure of composites. A satisfactory agreement with experiments reinforces further an adequacy of the results.

Nevertheless, the need for general principles, which would enable us to develop more rational and formal methods, cannot be overemphasized. Some of these are dealt with in the next section.

5.10 Rigorous Definitions of the Effective Response

In anticipation of the relations to be derived it will be convenient to introduce more condensed notation. To this end, following the previous section, we shall use boldface English letters to denote tensors. In particular, $s = s_{ij}$ (not σ_{ij}) will denote the linear stress tensor, $e = e_{ij}$ (not ε_{ij}) the linear strain tensor, and c the tensor of elastic moduli:

$$s = ce \quad . \tag{5.67}$$

In a multiphase medium c takes on value c_i in the i-phase.

Out basic assumption is that the composite at hand does possess an overall effective response, which may be identified by experiments with sufficiently large specimens. We denote this effective elastic tensor as C. Clearly, in view of the discussion on the concept of the RVE given in Section 5.5, the tensor C may be given the following sense:

$$C = \langle s \rangle / \langle e \rangle \quad , \tag{5.68}$$

where $\langle s \rangle$ and $\langle e \rangle$ are average stresses and strains.

Looking for rational definitions of the above average entities we should first ensure their very existence. To this end, imagine the boundary S of the specimen to be subject to homogeneous displacement

$$u_i = a_{ij}x_j \qquad \text{on} \qquad S \quad , \tag{5.69}$$

where a_{ij} are constants constituting symmetric matrix, $\boldsymbol{a}$. Then we get for the volume average strain

$$V^{-1}\int_V 1/2(u_{i,j}+u_{j,i})dv = V^{-1}\int_V a_{ij}dv = a_{ij} = \text{const.} \tag{5.70}$$

We may therefore specify the average strain appearing in (5.68) as the volume average, $\langle e\rangle = \boldsymbol{a}$, which has also clear mechanical sense, since it is consistent with the way of experimental determinations of this value. Accordingly, the average stress in (5.68) should then be also defined as the volume average

$$\langle s\rangle = V^{-1}\int_V s dv \quad . \tag{5.71}$$

The above procedure has an advantage of indicating the way to actually find the effective moduli, C. Namely, one has to subject the RVE to the boundary conditions given by (5.69) and then solve the elasticity problem with variable $\boldsymbol{c}$ to find the $\boldsymbol{s}$-field. Then (5.71) and (5.68) yield the overall moduli, C. Needless to say, this may turn out difficult to carry out, but it puts the concept of the effective behavior on the rational basis. Note that C is not, in general, equal to the volume average, $\langle c\rangle$.

Similarly, imagine the boundary S, to be subjected to homogeneous stresses, so that

$$s_{ij}n_j = b_{ij}n_j \qquad \text{on} \qquad S \quad , \tag{5.72}$$

where b_{ij} are constants and $\boldsymbol{n}$ is normal to S. The relation

$$V^{-1}\int_V s_{ij}dv = V^{-1}\int_S s_{ik}x_j n_k d\sigma$$

ensures the existence of the volume average, while (5.68) that of C. Clearly, this time, $\langle e\rangle$ follows from the solution to the proper boundary value problem.

A resort to the energy density, W, provides another look at the above definitions. With the help of (1.28) and on adjusting notations we obtain

$$\langle W\rangle = (2V)^{-1}\int_V s_{ij}u_{i,j}dv = (2V)^{-1}\int_S s_{ij}u_i n_j d\sigma \quad , \tag{5.73}$$

where use has been made of the equilibrium equation

$$s_{ij,j} = 0 \quad .$$

On substituting (5.69) into (5.73) we get

$$\langle W\rangle = (2V)^{-1}\int_V s_{ij}\langle e_{ij}\rangle dv = \langle s_{ij}\rangle\,\langle e_{ij}\rangle/2 = \langle e\rangle C\langle e\rangle/2, \tag{5.74}$$

since $a_{ij} = \langle e_{ij}\rangle$.

Thus, C is defined identically with the help of the mean strain and mean stress or the energy, which supports its validity. It makes also more plausible the conjecture that (5.68) describes the effective behavior for other than (5.69) or (5.72) boundary conditions. Definitions of overall properties could also be worked out on the basis of ensemble rather than spatial averaging, which will not be pursued here.

5.11 Bounds for Static Moduli

As has been noted earlier, the exact evaluation of the effective response of composites may turn out to be tedious at best in the case of ordered microstructure or unreasonable in the case of random microstructure. On the other hand, the accuracy of approximate solutions is often uncertain. Clearly, the availability of upper and lower bounds for C could in a radical way improve the situation. Furthermore, when these bounds are close enough to one another, they in fact resolve the problem of specification of the effective response by their very existence.

The simplest bounds follow from the classical variational principles of elasticity. Consider n-phase material and subject its boundary surface to the displacements given by (5.69). These may be continued in V and thought of as an admissible field in the context of the principle of minimum of potential energy described in Section 1.10. Then

$$W_{ex}(e) \leq \int_V ecedv/(2V) \quad , \tag{5.75}$$

where $W_{ex}(e)$ denotes the energy value provided by the exact solution and the integral term that provided by the above admissible field.

Invoking (5.69) as an admissible field and the indicator function, $f_i(r)$, introduced in Section 1.3 and defined by

$$f_i(r) = \begin{cases} 1, & r \text{ in phase } i \\ 0, & \text{otherwise} \end{cases}$$

we obtain for the average energy

$$\begin{aligned} \langle W \rangle &\leq a \sum_{i=1}^{n} \int_V c_i f_i(r) dva/(2V) \\ &= a \sum_{i=1}^{n} \phi_i c_i a/2 = \langle e \rangle C_v \langle e \rangle /2 \quad , \end{aligned} \tag{5.76}$$

where ϕ_i denotes the volume fraction of the i-phase, the Voight average, C_v, is

$$C_v = \sum_{i=1}^{n} \phi_i c_i \quad , \tag{5.77}$$

and use have been made of the relation

$$a = \langle e \rangle \quad .$$

Equations (5.74) and (5.76) show that $C_v - C$ is positive semidefinite and thus C_v provides the upper bound for components of C.

In a dual procedure we apply the traction boundary condition (5.72) and the complementary energy principle

$$W_{ex}(s) \le \int_V s c^{-1} s dv/(2V) \quad , \tag{5.78}$$

where $W_{ex}(s)$ is the exact energy value, while the integral yields its approximation following from an admissible self-equilibrated field. On continuing (5.72) in V we get

$$\langle W \rangle \le (2V)^{-1} \langle s \rangle \sum_{i=1}^{n} \int_V c_i^{-1} f_i(r) dv \langle s \rangle = \langle s \rangle \sum_{i=1}^{n} \phi_i c_i^{-1} \langle s \rangle /2 \quad . \tag{5.79}$$

Comparing this with the exact average energy

$$\langle W(s) \rangle = \langle s \rangle C^{-1} \langle s \rangle /2 \tag{5.80}$$

we obtain that $C_R^{-1} - C^{-1}$ is positive semidefinite, where the Reuss average, C_R is

$$C_R = \left[\sum_{i=1}^{n} \phi_i c_i^{-1} \right]^{-1} \quad . \tag{5.81}$$

Thus, (5.77) and (5.81) bound the effective moduli from above and from below, respectively. Unfortunately, these bounds are of practical significance only for small volume fractions and slight mismatch of elastic moduli of phases.

Far more stringent bounds may be derived with the help of new variational principles based on the notion of polarization tensor. For the purpose of discussion we present here the upper and lower bounds, which hold for a two-phase statistically isotropic composite

$$\begin{aligned} \frac{(K_1 + \tilde{K})\phi}{K_1 + \tilde{K} + (K_2 - K_1)(1-\phi)} &\le \frac{K_e - K_1}{K_2 - K_1} \\ &\le \frac{(K_1 + \tilde{\tilde{K}})\phi}{K_1 + \tilde{\tilde{K}} + (K_2 - K_1)(1-\phi)} \\ \frac{(\mu_1 + \tilde{\mu})\phi}{\mu_1 + \tilde{\mu} + (\mu_2 - \mu_1)(1-\phi)} &\le \frac{\mu_e - \mu_1}{\mu_2 - \mu_1} \\ &\le \frac{(\mu_1 + \tilde{\tilde{\mu}})\phi}{\mu_1 + \tilde{\tilde{\mu}} + (\mu_2 - \mu_1)(1-\phi)} \quad , \end{aligned} \tag{5.82}$$

where if $(\mu_2 - \mu_1)(K_2 - K_1) \ge 0$ then

$$\begin{aligned} &\tilde{K} = 4\mu_1/3, \quad \tilde{\tilde{K}} = 4\mu_2/3 \\ &\tilde{\mu} = \frac{3}{2}\left(\frac{1}{\mu_1} + \frac{10}{9K_1 + 8\mu_1}\right)^{-1}, \quad \tilde{\tilde{\mu}} = \frac{3}{2}\left(\frac{1}{\mu_2} + \frac{10}{9K_2 + 8\mu_2}\right)^{-1} \quad , \end{aligned} \tag{5.83}$$

while if $(\mu_2 - \mu_1)(K_2 - K_1) \leq 0$ then

$$\tilde{K} = 4\mu_2/3, \quad \tilde{\tilde{K}} = 4\mu_1/3$$
$$\tilde{\mu} = \frac{3}{2}\left(\frac{1}{\mu_1} + \frac{10}{9K_2 + 8\mu_1}\right)^{-1}, \quad \tilde{\tilde{\mu}} = \frac{3}{2}\left(\frac{1}{\mu_2} + \frac{10}{9K_1 + 8\mu_2}\right)^{-1} . \tag{5.84}$$

Like (5.77) and (5.81), these Hashin-Strikman bounds are valid for arbitrary phase geometry, but are much more restrictive, and in many cases are of practical usefulness. Comparison of the bounds for the bulk modulus, K_e, with (5.49), obtained via the composite spheres model, reveals that they are identical upon proper interpretation of the phases 1 and 2. Since (5.49) is exact for a particular phase geometry, this implies that the bounds for K_e are best possible in terms of volume fractions. This interesting fact has not yet been shown for the bounds for μ_e. Another remarkable feature of the above results is that the bounds coincide if $\mu_1 = \mu_2$, providing thus the exact solution for this case.

Now we may better appreciate the previously applied methods of evaluating the effective static response. It is clear from the above remarks that the composite sphere assemblage and thereby the effective scatterer approach do comply with (5.82). It can be shown that the differential scheme is also consistent with the bounds when applied to particular or fibre composites.

5.12 Wave Propagation in Random Composites

Long-wavelength waves in random elastic composites may often be investigated by the static methods given above. For these waves

$$a/\lambda \to 0 \quad , \tag{5.85}$$

where a is a typical microstructure dimension and λ is the shortest of the wavelengths involved. However, this approximation should be applied with great care. Strong dynamic effects may still take place even though this ratio is quite small, for example, in the case of rubberlike porous materials. Furthermore, if (5.85) does not hold, then a dynamic analysis appears necessary.

Usually random elastic composites exhibit explicit dispersive behavior, which is easily explained by the dependence of scattering cross section, $\gamma(\omega)$, on frequency. In general, wave propagation in such composites is influenced to a great extent by scattering. This may result in a quite surprising behavior.

In fact, scattering cross section, $\gamma(\omega)$, (see (5.20) and related equations) describes the amount of energy lost by an incident field because of its interaction with an inclusion. Since no compensating mechanism is available in view of random elastic microstructure the composite is seen by the incident wave as an attenuative medium, despite the conservative nature of the entire system. The destructive influence of irregularity on propagation of a disturbance was observed in other physical systems too. It follows, therefore, that the overall dynamic response may be conveniently described in terms of the complex wave number

$$k(i\omega) = \omega/c(\omega) + i\alpha(\omega) \quad , \tag{5.86}$$

where $c(\omega)$ is the phase speed, while $\alpha(\omega)$ the attenuation. This is similar to viscoelastic waves in a homogeneous medium considered in Section 2.15. In fact, for a plane attenuative dispersive wave we get

$$e^{ik(i\omega)x-i\omega t} = e^{-\alpha x}e^{i\omega(x/c-t)} \quad . \tag{5.87}$$

Note, that the notation for the argument, $i\omega$, in (5.86) merely indicates that the wave number is a complex value.

Specification of the attenuation, $\alpha(\omega)$, and the speed, $c(\omega)$, in terms of the microstructure is the main goal of the analysis. The concept of the RVE applies therefore again, similarly to the static case. It may appear that the presence of a typical microstructure dimension, like inclusion size, puts an upper limit on the wave frequency, in analogy with Debye's result mentioned in Section 1.15. It is not however the case due to randomness of the microstructure. In fact, a high-frequency wave will see the composite as more disordered than a low-frequency one, which makes such a bound irrelevant in the frameworks of the applied model.

It follows from the above that the elastic homogeneous medium cannot be a model for evaluation of an overall dynamic behavior of random elastic composites, for it does not show any dispersion, but a viscoelastic medium may be. This is the major difference compared to the static response treated earlier.

Instead of $\alpha(\omega)$ and $c(\omega)$ we may adopt the concept of dynamic effective modulus, $C(i\omega)$, or the compliance, $C^{-1}(i\omega)$. From the results of Section 2.14 we get the following interrelations between the two representations:

$$\begin{aligned} \operatorname{Re} C(i\omega) &= C_r = (Q_r^2 - Q_i^2)\rho_e \\ \operatorname{Im} C(i\omega) &= C_i = 2Q_r Q_i \rho_e \quad , \end{aligned} \tag{5.88}$$

where $Q(i\omega) = Q_r(\omega) + iQ_i(\omega)$ is the complex velocity defined by

$$Q(i\omega) = \frac{\omega}{k(i\omega)} = \left[\frac{C(i\omega)}{\rho_e}\right]^{1/2} \tag{5.89}$$

and ρ_e is the effective density of the composite to be discussed in more detail in Section 5.13. Equations (5.86) and (5.89) yield

$$\begin{aligned} \operatorname{Re} Q(i\omega) &= Q_r = c\omega^2/(\omega^2 + \alpha^2 c^2) \\ \operatorname{Im} Q(i\omega) &= Q_i = -\alpha\omega c^2/(\omega^2 + \alpha^2 c^2) \quad . \end{aligned} \tag{5.90}$$

Under the limit $\omega \to 0$ the modulus $C(i\omega)$, just defined, must be consistent with the proper static effective modulus, C,

$$\lim_{\omega \to 0} C(i\omega) = C \quad .$$

It should be pointed out again that $C(i\omega)$ has no sense as a local value, it describes the overall dynamic stiffness of a composite as a function of the microstructure and frequency.

As in the static case, the major difficulty is an adequate account for multiple interactions among inclusions. These may be divided into two kinds. First, the inclusions "influence" each other geometrically, in the sense that they should not overlap and their mutual positions are governed by certain probabilistic laws.

Second, there are purely dynamic interactions of their wave fields. Clearly, scattering by an inclusion in a pure matrix is different from that in a composite matrix, which has a modified impedance. In what follows we present some of the possible approaches to calculation of the effective dynamic response.

5.13 Causal Approach of Independent Scatterers

For the sake of certainty consider an elastic homogeneous isotropic matrix containing identical randomly dispersed inclusions. An incident plane wave gives first rise to primary scattered waves, which then cause secondary scattered waves, and so forth.

A natural way to begin the analysis of this tangled situation is to neglect rescattering, which is a reasonable approximation provided the volume fraction, ϕ, is small, and, perhaps, mismatch of the elastic moduli and that of the densities are slight. Under this assumption of independent scatterers the total power loss per unit volume due to scattering is, in view of (5.20), given by

$$-d\langle I\rangle = N\gamma\langle I\rangle dx \quad ,$$

where dx is the distance travelled by a disturbance, N is the number of inclusions per unit volume, and γ the scattering cross section of a single inclusion. This immediately yields the useful simple equation for decay of the average power flux

$$\langle I\rangle = \langle I_o\rangle e^{-N\gamma x} \quad . \tag{5.91}$$

Since $\langle I\rangle$ is quadratic in the wave amplitude, it follows from (5.87) that

$$\alpha(\omega) = N\gamma(\omega)/2 \quad , \tag{5.92}$$

which expresses the attenuation in terms of the microstructural parameters, N and γ.

Hence, in this approximation, the problem reduces to determination of the wave velocity, $c(\omega)$. To this end, we appeal to linearity, causality, and passivity of the effective medium to be constructed. In fact, this should comply with the above constraints to represent a physically meaningful model, which enables us to invoke the $K-K$ relations, for example, those given by (2.140). Since the refractive index, $n(i\omega)$ is

$$n(i\omega) = \tilde{c}k(i\omega)/\omega \quad ,$$

we put the $\tilde{c} = c_o$ as a reference speed, and rewrite the first of (2.140) as

$$c(\omega) = c_o\left[1 + \frac{2\omega^2 c_o}{\pi} P\int_0^\infty \frac{\alpha(\omega')d\omega'}{\omega'^2(\omega'^2-\omega^2)}\right]^{-1} \quad , \tag{5.93}$$

where c_o is the static limit,

$$c_o = \lim_{\omega\to 0} c(\omega) \quad . \tag{5.94}$$

Clearly, (5.92) and (5.93) provide the solution, if the static limit, c_o, is known too. This poses no problem, since

$$c_o^2 = \lim_{\omega \to 0} [C(i\omega)/\rho_e] = C/\rho_e \quad , \tag{5.95}$$

which can be determined from an independent static analysis. In the case of solid media, the effective static density, ρ_e is given by (5.47) as

$$\rho_e = \phi\rho_2 + (1-\phi)\rho_1 \quad , \tag{5.96}$$

which coincides with the gravitational density.

It is to be noted that the K–K relations may also be written for the effective dynamic modulus, $C(i\omega)$. In fact, the effective medium is thought of as a viscoelastic and, say, (1.47) apply upon associating $C(i\omega)$ with $G(i\omega)$. This is in accord with the K–K relations, which are valid for any linear causal and passive system, regardless of its nature or a particular attenuation mechanism involved. It means, among others, that the physical nature of the overall losses does not matter as far as the wave speed is found via these relations. Following the analogy with a viscoelastic homogeneous medium, we may assume that the effective density does not depend on frequency and thus (5.96) extends to a dynamic case. Indeed, if ρ_e were frequency-dependent, then this dependence should be of quite specific form to insure that the other related values, $C(i\omega)$ and $k(i\omega)$, satisfy the K–K relations too. Equation (5.96) thus defines the effective density of the RVE and is inapplicable to smaller volumes, for which dependence on frequency may occur. We adopt therefore (5.96) as a reasonable approximation for the dynamic case, valid in the frameworks of the applied model, which assumes no knowledge of statistical information of higher order.

We conclude, that upon omission of the multiple scattering effects the solution is given by (5.92) – (5.96), where the microstructure comes into play solely through scattering cross-section of a single inclusion, γ, number of inclusions per unit volume, N, and the static limit, c_o. The approach is thus extremely modest in its requirements to the microstructure description.

A remarkable feature of the treatment is that it puts no frequency limitations, the effective response may be evaluated for the entire frequency interval, $0 \le \omega < \infty$. In fact, subtracting (2.140), we get, that the static limit, c_o, and the geometric limit, c_∞,

$$c_\infty = \lim_{\omega \to \infty} c(\omega) \tag{5.97}$$

are related by

$$c_\infty/c_o = \left[1-(2c_o/\pi)\int_0^\infty \alpha(\omega)\omega^{-2}d\omega\right]^{-1} \quad . \tag{5.98}$$

The physical sense of this limit becomes clear if we recall the concept of signal (shock) velocity discussed in Sections 1.16 and 2.15. Since $\alpha(\omega) > 0$ by definition, this implies

$$c_\infty > c_o \quad , \tag{5.99}$$

which means, among others, that the lower bound for c_o holds for c_∞ too. A point to note is that this approach is solely based on simple energy considerations and appeal to such physically meaningful features of the effective medium as its causality, passivity, and linearity.

5.14 Causal Differential Media

It has been noted in the end of Section 5.12, that multiple interactions may be classified as either geometric or dynamic. We must introduce statistical information of higher order, which deals with correlations among inclusion locations, to account for the former, and know how to modify the impedance of the medium surrounding an inclusion to account for the latter. Clearly, the geometric correlations are of importance for appreciable volume fractions, while the dynamic interactions may be significant even for dilute mixtures if scattering is intensive enough. We present in what follows an approximate approach, which takes into account multiple scattering effects due to the dynamic interactions only.

To this end we invoke the concept of differential effective media introduced in Section 5.8 and extend it to a dynamic case by making use of the approach of independent scatterers as a building block. Like its static precursor, the dynamic differential scheme is thought of as a realizable model of composites, but it differs from it in many aspects, which are discussed below.

Consider a two-phase solid mixture with a lossless homogeneous background medium and randomly dispersed identical lossless inclusions at the volume concentration ϕ. We begin the construction process with a matrix of phase 1 and imbedded inclusions of phase 2 in dilute concentration, which is chosen so as to ensure slight rescattering. A care also must be taken to avoid overlapping and a consequent occurrence of inclusion-size distribution. This aspect will be discussed in some detail later. Next, a homogenization is carried out by employing a theory of the effective dynamic response, and then new inclusions are imbedded into the "homogeneous" matrix. By repeating this incremental procedure one arrives at the volume concentration prescribed.

Let $\{I\}$ be a set of parameters which identify relevant physical properties of inclusions and $\{M\}$ that of a background medium. Clearly, in the above construction process $\{I\}$ remains constant, while $\{M\}$ takes current values defined by homogenization. This differs from the static differential scheme for which $\{M\}$ consists of the static moduli solely. In the rest of this section tilda denotes a current value.

The relation between the increase in total absolute concentration, $d\phi$, and the current increase, $d\tilde{\phi}$, is given by (5.50), while the evolution law of a static modulus, C, should be clear from the example given in Section 5.8. Therefore, we go over to the remaining parameters of the set $\{M\}$, namely, to the mass density, $\tilde{\rho}_1$, the attenuation, $\tilde{\alpha}_1(\omega)$, and the speed, $\tilde{c}_1(\omega)$.

For the case of solid composite media, the effective density, ρ_e, is linear in ϕ (see (5.47)), which provides

$$\tilde{\rho}_1 = (1-\phi)\rho_1 + \phi\rho_2 \quad . \tag{5.100}$$

As has been noted previously (5.100) should be viewed as a reasonable approximation, in particular, when no statistical information of higher order, such as a pair-correlation function, is available, a restriction mentioned previously.

Now we turn to the attenuation, $\alpha(\omega)$, which under the absence of rescattering is given by (5.92)

$$\alpha(\omega) = N\gamma/2 = \phi\gamma/(2V) \quad , \tag{5.101}$$

where V is the inclusion volume. This approximate additive rule can be naturally adapted in the frameworks of the differential scheme.

Assume that the total number of inclusions, N, is represented as

$$N = N_1 + N_2 + \dots \quad ,$$

where each of N_i is small enough to allow for omission of multiple scattering effects. At the first step we insert N_1 inclusions, then apply the approach of independent scatterers, and find the "first" homogeneous medium, which is attenuative and dispersive. Then we insert N_2 inclusions and repeat the procedure, while taking into account the above behavior of the matrix. To this end, we apply again incremental removal-replacement process leading to a realizable effective medium, which has been described in Section 5.8. Namely, at each infinitesimal step the volume of "homogeneous" material is removed by $\Delta\tilde{\phi}$ and replaced with the same volume of inclusion material. Thus, in the first approximation,

$$\tilde{\alpha}(\phi + \Delta\phi) \approx \tilde{\alpha}(\phi)(1 - \Delta\tilde{\phi}) + \Delta\tilde{\phi}\tilde{\gamma}/(2V) \quad , \tag{5.102}$$

where $\tilde{\gamma}$ is current scattering cross section of a single inclusion in homogenized matrix and $\Delta\tilde{\phi}$ the increase in the current concentration. Since the homogenized matrix is a dispersive medium, the frequency-dependence of its elastic moduli must be taken into account in computing the elastic scattering cross-section, $\tilde{\gamma}$. So, there is a coupling between the two terms in the right-hand side of (5.102). The first term describes the "accumulated" attenuation of the matrix while the second scattering losses due to the current increase, $\Delta\tilde{\phi}$. Equation (5.102) thus generalizes the well-known additive law for attenuation due to different independent mechanisms by taking into account the influence of the first term on the second. Equations (5.102) and (5.50) yield the following evolution law for attenuation

$$\frac{d\tilde{\alpha}}{d\phi} = \frac{\tilde{\gamma}/(2V) - \tilde{\alpha}}{1 - \phi} \quad , \tag{5.103}$$

with initial condition $\tilde{\alpha} = 0$ at $\phi = 0$. Note that a more accurate treatment should invoke the concept of scattering losses in a viscoelastic matrix, for which the reader should consult the comments on Problem 5-4.

It remains to find the phase velocity, $\tilde{c}(\omega)$, to complete $\{M\}$. On invoking the K–K relation we get

$$c(\tilde{\omega}) = \tilde{c}_o\left[1 + \frac{2\omega^2}{\pi}\tilde{c}_o P\int_0^\infty \frac{\tilde{\alpha}(\omega')d\omega'}{\omega'^2(\omega'^2 - \omega^2)}\right]^{-1} \quad . \tag{5.104}$$

This build-up process terminates upon arrival at the volume fraction prescribed. The model just constructed, which is referred to as the causal differential medium, is a natural extension of the approach of independent scatterers.

The model enables us to find the response for the entire frequency interval $0 \leq \omega < \infty$, including c_∞, which, according to (5.98), is given by

$$\tilde{c}_\infty/\tilde{c}_o = \left[1-(2\tilde{c}_o/\pi)\int_0^\infty \tilde{\alpha}(\omega)\omega^{-2}d\omega\right]^{-1} .$$

It has been shown in Sections 1.16 and 2.15 that c_∞ is the speed of propagation of a small-amplitude shock. Hence, the approach provides also an explicit information concerning the transient response.

Having obtained $c(\omega)$ and $\alpha(\omega)$ we may derive the expression for the modulus, $C(i\omega)$, via (5.96), (5.90) and (5.88). In particular, in view of (5.24), (5.92) and (5.97) we get for the complex wave speed

$$\lim_{\omega\to\infty} \operatorname{Re} Q(i\omega) = c_\infty$$

$$\lim_{\omega\to\infty} \operatorname{Im} Q(i\omega) = 0 .$$

Then (5.88) yields the shock modulus, C_∞,

$$C_\infty = \lim_{\omega\to\infty} C(i\omega) = c_\infty \rho_e ,$$

which describes the transient instantaneous stiffness of a random elastic composite.

Like the static version, the approach disregards the geometrical correlations and involves therefore no higher order information concerning the phase geometry. The above build-up procedure can be fulfilled in the two ways, which lead either to "overlapping" or "stirred together" inclusions. In the latter version a care is taken to place new inclusions so as to avoid mutual "penetration" of scatterers. Equations describing the static response (see, for example, (5.55)) hold in either case since they incorporate no information concerning the inclusion size or a correlation function. On the other hand, in the dynamic case the response does show sensitivity to the size of inclusions. Thus, to arrive at the composite under consideration, which contains identical inclusions, a care should be taken to place new scatterers so as to avoid overlapping and strong rescattering. This may put a limitation on the admissible concentration, ϕ, a restriction mentioned previously. A more complete treatment of this aspect appears possible upon accounting for geometrical correlations.

In problems of stochastic nature it is typically difficult to evaluate rigorously the accuracy of the solution. This may be achieved by comparison rather than by an absolute estimate. In this sense, the approach of causal differential media may be useful for evaluation of the less accurate technique of independent scatterers. In turn, a clearer appreciation of the former may be possible when a more accurate theory becomes available.

In the next section we consider an example, which enables us to clarify and compare the presented techniques as well as to discover a pattern of the dispersion curves, $\alpha(\omega)$ and $c(\omega)$, typical of random elastic composites.

5.15 Waves in Fibre Composites. Typical Dispersion Curves

We consider axial shear waves in random fibre composites. Besides its practical relevance, this problem has always served as a convenient "touchstone" for evaluation of basic effects because of its scalar nature.

The composite contains identical fibers with radius a, which are aligned along the z-axis and randomly placed in the xy-plane (Figure 5-8). Denote the shear modulus and the density of the matrix as μ_1 and ρ_1 and those of fibers as μ_2 and ρ_2, respectively. The volume (area) fraction, ϕ, is the only information available concerning the phase geometry and is given by

$$\phi = N\pi a^2 \quad ,$$

with N the number of inclusions per unit volume (area). The axially polarized shear wave of displacement, u_z, may be written as

$$u_z = e^{i[k(i\omega)x-\omega t]} = e^{-\alpha(\omega)x} e^{i\omega[x/c(\omega)-t]}$$

where $\alpha(\omega)$ and $c(\omega)$ are to be specified in terms of the microstructure.

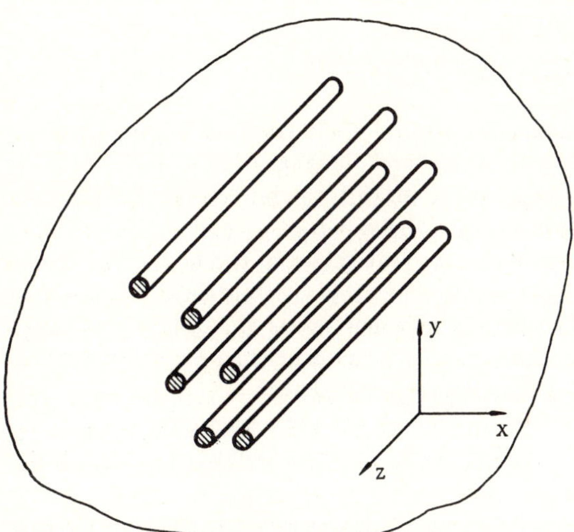

Fig. 5-8 Random fibre-reinforced composite.

We apply the approach of causal differential media. The tensor of effective static moduli, C, reduces to the shear axial modulus, μ_e. Then (5.52) yields its current value, while the current density is given by (5.100). Consequently, the current static limit, $\tilde{c}_o$, appearing in (5.104) is

$$\tilde{c}_o^2 = \tilde{\mu}_e/\tilde{\rho}_1 \quad . \tag{5.105}$$

It remains to specify the current attenuation, $\tilde{\alpha}(\omega)$, which is closely related to scattering cross-section, $\tilde{\gamma}(\omega)$. We turn therefore to interaction of a cylinder with the SH-waves investigated in Sections 5.1 and 5.3. The general expression for γ

is given by (5.27), where the coefficients A_n, $n = 0, 1, 2...$ are given by (5.9). On replacing the values associated with the matrix by the current ones, we get

$$\tilde{\gamma} = 2\Big(2|\tilde{A}_o|^2 + \sum_{n=1}^{\infty} |\tilde{A}_n|^2\Big)/\tilde{k}_1 \quad , \tag{5.106}$$

with

$$\tilde{A}_n = -i^n \xi_n \frac{\mu_2 k_2 J_n(\tilde{q}_1) J_n'(q_2) - \tilde{\mu}_1 \tilde{k}_1 J_n'(\tilde{q}_1) J_n(q_2)}{\mu_2 k_2 H_n(\tilde{q}_1) J_n'(q_2) - \tilde{\mu}_1 \tilde{k}_1 H_n'(\tilde{q}_1) J_n(q_2)} \quad . \tag{5.107}$$

Here $\tilde{q}_1 = \tilde{k}_1 a$ and $q_2 = k_2 a$ are the real dimensionless wave numbers,

$$\begin{aligned} \tilde{q}_1 &= \omega a / \tilde{c}_1(\omega) = \omega a / (\tilde{\mu}_1/\tilde{\rho}_1)^{1/2} \\ q_2 &= \omega a / c_2 = \omega a / (\mu_2/\rho_2)^{1/2} \end{aligned} \tag{5.108}$$

and

$$\xi_n = \begin{cases} 1, & n = 0 \\ 2, & n \leq 1 \end{cases} \quad .$$

The coupled equations (5.101) – (5.107) can be best solved by simulating numerically the above step-by-step incremental procedure of constructing the final phase geometry. Only (5.100), which immediately yields the effective density $\tilde{\rho}_1$, may not need this treatment.

On stipulating initial small $\phi = \Delta\tilde{\phi}$, (5.101), (5.100) and (5.55) are first evaluated, which provide $\tilde{\alpha}_1$, $\tilde{\rho}_1$, and $\tilde{\mu}_1$ of the "first" homogenized matrix. This step is completed by computing $\tilde{c}_o$ and $\tilde{c}_1(\omega)$ from (5.105) and (5.104). For the next step, when ϕ gets a new small increment, the values of μ_1 and k_1 appearing in (5.106) and (5.107) are replaced by the current values of the homogenized matrix, $\tilde{\mu}_1(\omega)$ and $\tilde{k}_1(\omega)$, respectively, which were previously obtained. On invoking (5.100), (5.101), (5.105), and (5.104) one finds new values of $\tilde{\rho}_1$, $\tilde{\alpha}_1(\omega)$ and $\tilde{c}_1(\omega)$. This simulation of constructing a more concentrated mixture for a stipulated ω continues until the prescribed volume fraction is reached.

The number of steps needed, n, depends on the convergence rate of the process. Clearly, for $n = 1$ the process reduces to the approximation of independent scatterers.

Figure 5-9 shows the attenuation and the wave velocity in boron-aluminum composite with the parameters $\rho_1/\rho_2 = 1.075$, $\mu_1/\mu_2 = 0.155$, and the volume fraction $\phi = 0.185$. It is seen that the effects of multiple scattering taken into account in the model decrease the first peak and shift it to a higher frequency.

The dispersion curves, $\alpha(\omega)$ and $c(\omega)$, found in the above problem are quite typical of random composites. These curves show a pattern of behavior independent to some degree of the microstructure. Namely, $\alpha(\omega)$ increases first with increasing frequency in agreement with the Rayleigh law for $\gamma(\omega)$ given by (5.22), then there may be a resonance overshoot due to resonance scattering, and then stabilization in the region of high frequencies. The associated $c(\omega)$, which is related to $\alpha(\omega)$ through a singular integral, decreases at first with increasing frequency, but it may then rise to quite a large value, if the resonance in $\alpha(\omega)$ is strong enough. In the extreme case $c(\omega) \to \infty$ in the vicinity of resonance, which

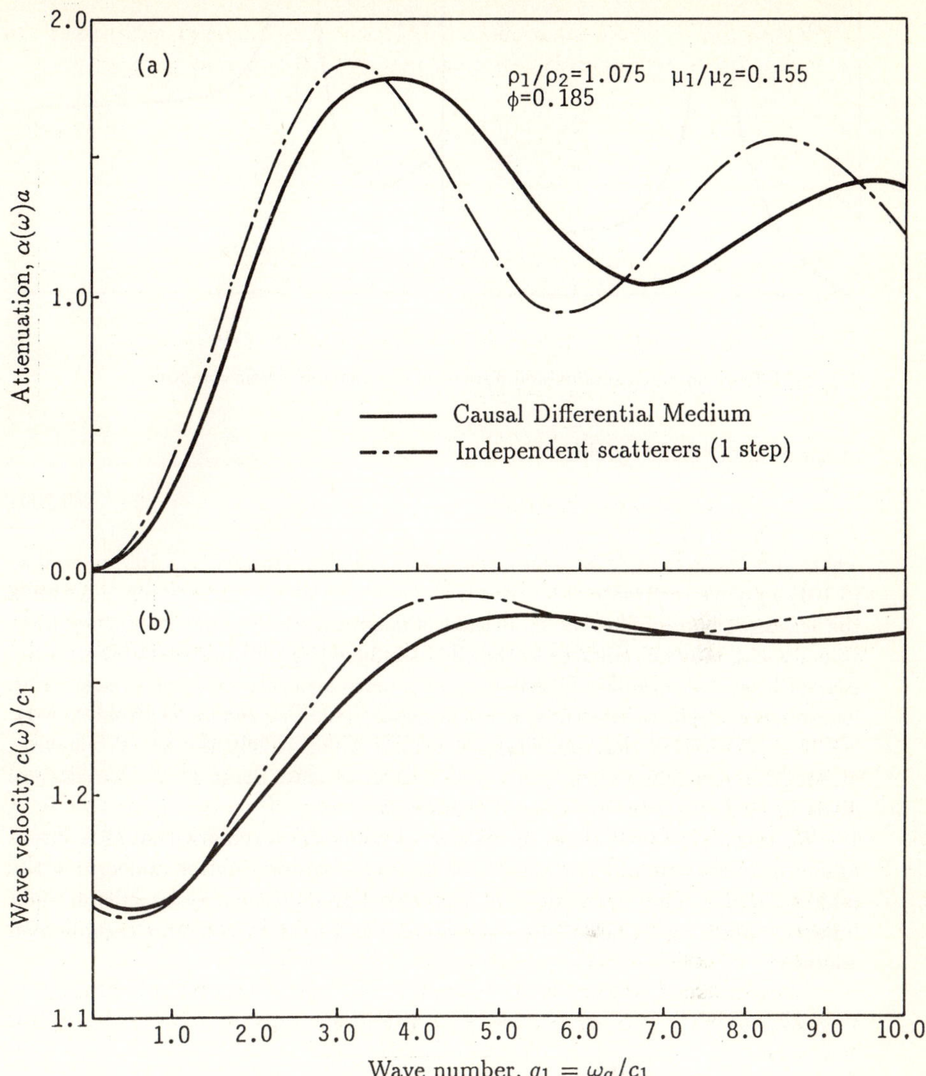

Fig. 5-9 Attenuation and wave velocity for boron-aluminum composite.

means that the medium may not support running waves because of the over-attenuation and incoherent motions. The associated frequency interval is called the stop-band. When $\omega \to \infty$ the $c(\omega)$-curve approaches a constant value $c_\infty > c_o$, as given by (5.98) and (5.99). This resonance behavior is observed, for example, for porous rubberlike materials and is sketched in Figure 5-10.

A relevant point to note is that (5.103) predicts a slight effect of multiple scattering on attenuation not only for low but for high frequencies too. In fact, for $\omega \to \infty$, γ is given by (5.24) regardless of the material characteristics. Then

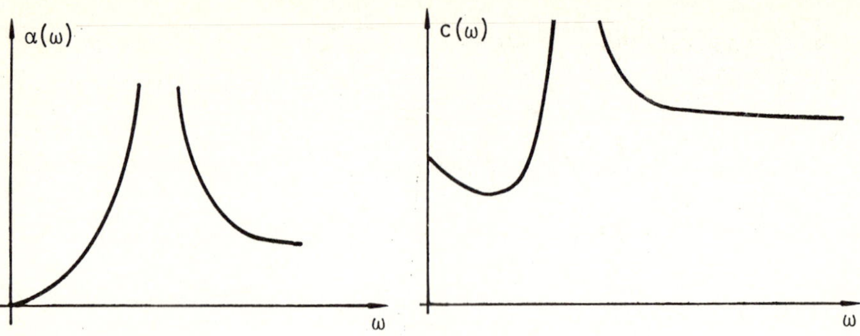

Fig. 5-10 Resonance attenuation and dispersion in a random elastic composite.

integration of (5.103) yields

$$\lim_{\omega\to\infty} \alpha(\omega) = \alpha_\infty = \phi A/V \tag{5.109}$$

where A is the geometric cross-section of an inclusion. However, (5.109) as well as (5.101) may overestimate the losses if there is constructive interference between the scattered and incident fields. Also, the magnitude of the high-frequency wave velocity, c_∞, may be sensitive to the effects induced by multiple scattering. In the case of large "accumulated" attenuation a more accurate analysis of scattering losses in a viscoelastic matrix appears necessary (see Comments on Problem 5-4).

It is of interest that polycrystals exhibit qualitatively the same behavior, although a resonance overshoot is, of course, absent. Some of the details are given in Problem 5-14 and related comments.

We complete the analysis of disordered media by a remark that it is inseparable from the amount and nature of the available information concerning the microstructure. The "resolution" of theoretical predictions may be substantially improved upon a finer description of materials and a resort to a more complicated analysis.

5.16 Waves in Ordered Systems. Atomic Lattice

Unlike random media, investigations of periodic systems involve no uncertainty concerning the microstructure, which is supposed to be completely given. Accordingly, the approach resembles to a considerable extent that of the classical elasticity, and the exact solution may be pursued. A proper formalism may be developed by the theory of differential equations with periodic coefficients. However, it appears more attractive to investigate their properties by considering some of the typical problems.

We treat below three problems dealing with dynamics of one-dimensional atomic chains and of a layered composite, which illustrate in a simple way the main features of waves propagation in periodic systems.

5.16.1 Primitive Lattice

A simple one-dimensional monatomic lattice of identical particles of mass m spaced at intervals a and connected by springs is shown in Figure 5-11. Because of spatial variability of the density and stiffness, the lattice represents a particular case of ordered inhomogeneous media.

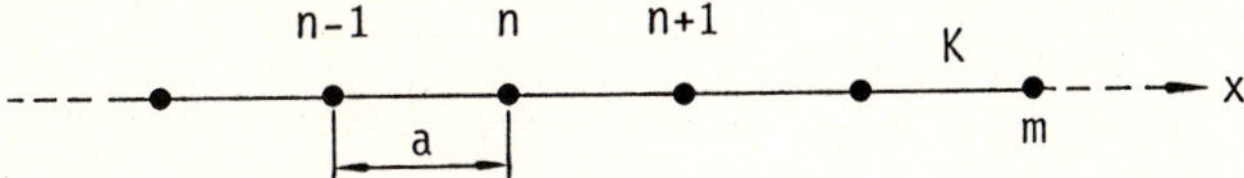

Fig. 5-11 Primitive atomic lattice.

Let u_n be the displacement of the n-th mass from its equilibrium position and F_n the associated restoring force. Then the equation of motion is

$$F_n = K(u_{n+1} - u_n) - K(u_n - u_{n-1}) = m\ddot{u}_n \tag{5.110}$$

where K is the spring constant.

When a solution of the running wave

$$u_n = Ae^{i(kna-\omega t)}, \qquad u_{n\pm 1} = u_n e^{\pm ika}$$

is substituted in (5.110) we get the dispersion relation

$$\omega^2 = 4K\sin^2(ka/2)/m \tag{5.111}$$

or

$$\omega = \pm 2\sin(ka/2)(K/m)^{1/2} \quad , \tag{5.112}$$

where sign may be chosen so as to make frequency positive.

The characteristic feature of (5.112) is its periodicity,

$$\omega(k + 2\pi/a) = \omega(k) \quad ,$$

which is obviously a manifestation of the ordered microstructure. Figure 5-12, which is computed from (5.112), shows that $\omega(k)$ is a repetition of its variation in the interval

$$-\pi/a < k \le \pi/a \quad ,$$

which is called the first Brillouin zone. Accordingly,

$$-2\pi/a < k \le -\pi/a, \qquad \pi/a < k \le 2\pi/a$$

constitutes the second zone, and so forth.

Besides periodicity, (5.112) shows the presence of the upper frequency limit, ω_+,

$$\omega \le \omega_+ = 2(K/m)^{1/2}$$

which is another typical feature of ordered systems. This has been noted first in Section 1.15 in a discussion of Debye's frequency.

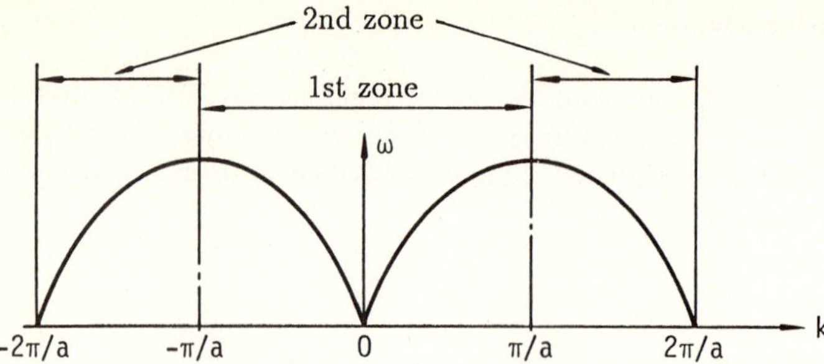

Fig. 5-12 Dispersion relations for a primitive lattice.

The phase velocity is

$$c = \omega/k = 2(K/m)^{1/2}\sin(ka/2)/k \quad . \tag{5.113}$$

As $ka \to 0$ with k fixed, the lattice becomes more "dense" and the foregoing relation reads

$$\rho c^2 = Ka = \text{const} \quad ,$$

where $\rho = m/a$ may be called the linear density. Hence, the lattice is seen by the wave as a homogeneous medium, which now exhibits no dispersion.

Needless to say, due to an extremely simplified structure, the monatomic lattice cannot possess all the main features of wave propagation in periodic systems. We consider therefore a more complicated case.

5.16.2 Diatomic Lattice

Diatomic chain comprises two kinds of particles whose masses are alternately M and m, connected by identical springs (Figure 5-13). The masses, M and m, are in equilibrium at $x = (2n+1)a$ and $x = 2na$, respectively. The equations of motion are

$$\begin{aligned} m\ddot{u}_{2n} &= K(u_{2n+1} - 2n_{2n} + u_{2n-1}) \\ M\ddot{u}_{2n+1} &= K(u_{2n+2} - 2u_{2n+1} + u_{2n}) \quad . \end{aligned} \tag{5.114}$$

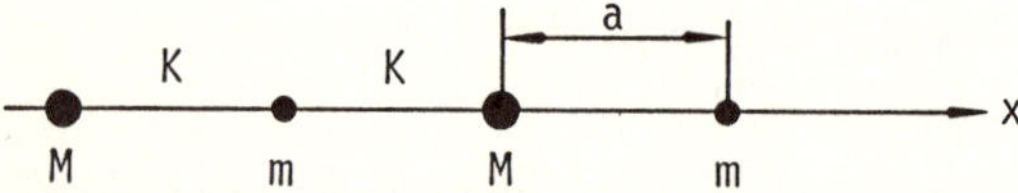

Fig. 5-13 Diatomic lattice.

Substituting

$$u_j = A_j e^{i(jka - \omega t)} \quad ,$$

for u we get the system of homogeneous equations

$$\begin{bmatrix} (2K - m\omega^2) & -2K\cos ka \\ -2K\cos ka & (2K - M\omega^2) \end{bmatrix} \begin{Bmatrix} A_{2n} \\ A_{2n+1} \end{Bmatrix} = 0$$

which, in turn, provides the determinantal equation

$$\omega^2 = Kq \pm K[q^2 - 4\sin^2(ka)/(Mm)]^{1/2} \quad , \tag{5.115}$$

with

$$q = (M+m)/(Mm) \quad .$$

Thus, the function $\omega(k)$ is again periodic. However, this time it has two branches according to the chosen sign in (5.115). Figure 5-14, which depicts the situation, is instructive. It shows that for the lower branch

$$0 \leq \omega^2 \leq 2K/M \quad ,$$

whereas for the upper

$$2K(m^{-1} + M^{-1}) \geq \omega^2 \geq 2K/m \quad ,$$

when $0 < ka < \pi/2$. These are known as the acoustic and optical branches, respectively.

For the lower branch the low-frequency expansion is given by

$$\omega_{ac}^2 \simeq 2K(ka)^2/(M+m), \quad ka \ll 1 \quad ,$$

whereas for the upper one by

$$\omega_{opt}^2 \simeq 2K(M+m)/(Mm), \quad ka \ll 1 \quad .$$

We observe that, to this approximation, the lower branch indeed behaves like a sound wave, with phase velocity $c = \omega/k = \text{const}$, which can also be shown to be the group velocity. The amplitudes A_{2n} and A_{2n+1} are equal and have the same sign.

Unlike this, the upper branch has zero group velocity,

$$c_g = \partial\omega_{opt}/\partial k = 0$$

and the atomic displacements are in antiphase, constituting thus an internal non-propagating mode of motion. These vibrations of a crystal are most sensitive to the presence of electromagnetic fields, which explains their name. The long wavelength group velocities of the above modes are shown as tangent to the two branches at $k = 0$ in Figure 5-14.

Another effect is the presence of a stop-band. In fact, in the interval

$$2K/M < \omega^2 < 2K/m$$

no real wave number exists. The solution given by (5.115) breaks down in this interval, indicating that no stationary running waves may propagate through the lattice. The appearance of a stop-band is a further manifestation of the periodic character of the structure, which has occurred already in the primitive lattice in the form of the upper frequency bound. Note that a stop-band may also appear in the frequency spectrum of random elastic composites. But the mechanism responsible in that case is resonant scattering and destructive phase interference among the waves involved, which results in over-attenuation, in contrast to the system at hand. In fact, a lattice with chaotically placed atoms also shows an attenuative response.

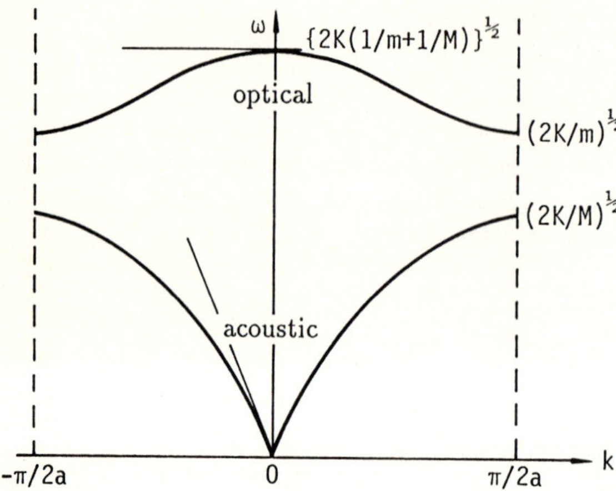

Fig. 5-14 Dispersion relations for a diatomic lattice.

5.17 Waves in Layered Composites

The layered structure shown in Figure 5-15 provides a further example of a periodic system. The medium consists of alternating layers of two different elastic materials, with the elastic constants, λ_n and μ_n, and the densities, ρ_n, where $n = 1, 2$.

Consider the plane P-wave propagating in the direction normal to the laminates. The displacement in the n-th layer, u_n, may be represented by

$$u_n(x,t) = U_n(x)e^{i(kx-\omega t)}, \quad n = 1,2 \tag{5.116}$$

where the function, $U_n(x)$, and the wave number, $k = \omega/c$, are unknown. However, due to the periodicity of the system (see Figure 5-15), we get

$$U_n(x+h) = U_n(x), \quad n = 1,2 \tag{5.117}$$

and

$$U_1(x = h_1) = U_2(x = -h_2) \quad . \tag{5.118}$$

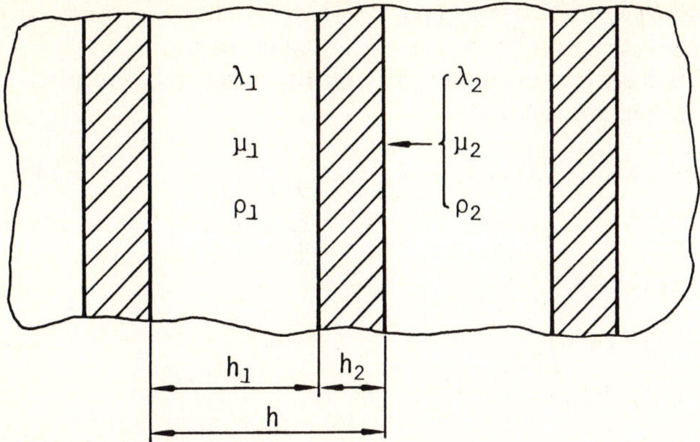

Fig. 5-15 Layered composite under consideration.

The same conditions can also be written for the stress.

Furthermore, the stress and displacement are continuous at the interface, say $x = 0$, since the laminates are supposed to be firmly bonded. Thus,

$$\begin{aligned} U_1(x=0) &= U_2(x=0) \\ \sigma_{xx1}(x=0) &= \sigma_{xx2}(x=0) \quad , \end{aligned} \tag{5.119}$$

where the third subscript denotes a proper layer.

The above relations enable one to determine the unknown values. Substitution of (5.116) into the governing equations given by

$$(\lambda_n + 2\mu_n)u_{n,xx} = \rho_n \ddot{u}_n, \quad n = 1,2$$

shows that

$$U_n(x) = A_n e^{i\delta_{an}x} + B_n e^{i\delta_{bn}x}, \quad n = 1,2 \quad ,$$

with

$$\begin{aligned} \delta_{an} &= \omega(1/c_n - 1/c) \\ \delta_{bn} &= -\omega(1/c_n + 1/c) \quad . \end{aligned}$$

Here A_n, B_n, and c are unknown values, whereas ω and $c_n^2 = (\lambda_n + 2\mu_n)/\rho_n$ are given.

Invoking the above periodicity conditions we get the four equations

$$\begin{aligned} & A_1 e^{i\omega h_1 \delta_{a1}} + B_1 e^{i\omega h_1 \delta_{b1}} = A_2 e^{-i\omega h_2 \delta_{a2}} + B_2 e^{-i\omega h_2 \delta_{b2}}, \\ & \frac{(\lambda_1 + 2\mu_1)}{c_1}\left[A_1 e^{i\omega h_1 \delta_{a1}} + B_2 e^{i\omega h_1 \delta_{b1}}\right] \\ & \qquad = \frac{(\lambda_2 + 2\mu_2)}{c_2}\left[A_2 e^{-i\omega h_2 \delta_{a2}} + B_2 e^{-i\omega h_2 \delta_{b2}}\right], \\ & A_1 + B_1 = A_2 + B_2, \end{aligned}$$

$$\frac{(\lambda_1 + 2\mu_1)}{c_1}(A_1 + B_1) = \frac{(\lambda_2 + 2\mu_2)}{c_2}(A_2 + B_2) \quad .$$

The determinant of this system must vanish to ensure a nontrivial solution. This gives the periodic dispersion relation,

$$\cos(h\omega/c) = \cos\lambda_1 \cos\lambda_2 - \Delta \sin\lambda_1 \sin\lambda_2 \tag{5.120}$$

with

$$\lambda_n = \omega h_n/c_n, \quad n = 1,2$$

$$\Delta = 0.5[\rho_1(\lambda_1 + 2\mu_1) + \rho_2(\lambda_2 + 2\mu_2)] / [\rho_1\rho_2(\lambda_1 + 2\mu_1)(\lambda_2 + 2\mu_2)]^{1/2} \quad .$$

In terms of the wave number, k, this relation can be rewritten as

$$\cos kh = \cos\frac{kh_1c}{c_1}\cos\frac{kh_2c}{c_2} - \Delta\sin\frac{kh_1c}{c_1}\sin\frac{kh_2c}{c_2} \quad . \tag{5.121}$$

Note, that according to this equation, the wave transmission occurs only for $|\cos kh| \leq 1$, while the condition $|\cos kh| > 1$ defines a stop-band.

A convenient graphical illustration follows for a particular case given by

$$h_1/c_1 = h_2/c_2 = q, \quad \rho_1 = \rho_2 \quad .$$

Then (5.121) reads

$$\cos kh = 1 - \sin^2(\omega q)\frac{(c_1 + c_2)^2}{2c_1c_2} \quad . \tag{5.122}$$

Equation (5.122) is shown in Figure 5-16, where the hatched areas correspond to the stop bands. This displays filtering properties of the layered composite under consideration.

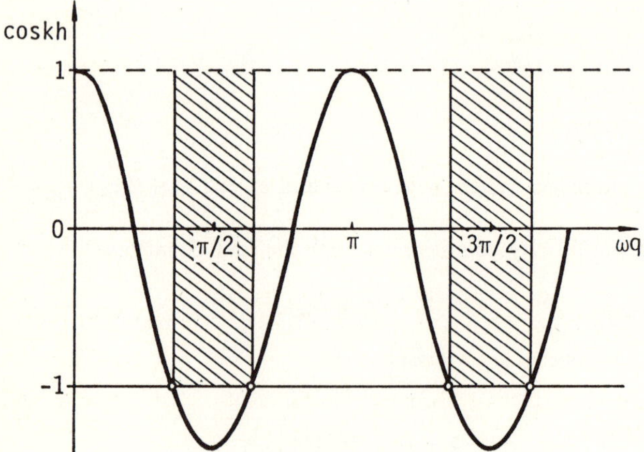

Fig. 5-16 Dispersion relation for a layered composite.

Knowledge of the dispersion relation, $k = k(\omega)$, paves a way for calculation of a transient response. By going over to the complex wave number, $k(i\omega)$, whose real part is governed by (5.121) and whose imaginary part reflects either viscoelastic or scattering losses within the layers, we may investigate how the periodic structure and attenuation influence the pulse propagation. A possible approximate approach is outlined in the comments on Problem 5-15.

The static response follows from the low-frequency expansion of (5.120)

$$h^2/c_o^2 = h_1^2/c_1^2 + h_2^2/c_2^2 + 2\Delta h_1 h_2/(c_1 c_2)$$

where c_o is the static limit. On the other hand,

$$c_o^2 = (\lambda + 2\mu)_e/\rho_e \quad , \tag{5.123}$$

where the effective static parameters of the composite have been introduced. Setting, in agreement with (5.96),

$$\rho_e = (h_1\rho_1 + h_2\rho_2)/h \quad , \tag{5.124}$$

we get

$$(\lambda + 2\mu)_e = \frac{h}{h_1/(\lambda_1 + 2\mu_1) + h_2/(\lambda_2 + 2\mu_2)} \quad , \tag{5.125}$$

which may also be derived in a straightforward way by considering a one-dimensional deformation of the composite.

Problems

5-1 Deduce the scattered field of a long cylindrical cavity subjected to SH-waves.

5-2 Solve the elastostatic problem by

$$\nabla^2 u_z(r,\theta) = 0$$

$$\sigma_{\theta z} = \mu u_{z,\theta}/r = -\sigma_{\theta z}^i \sin\theta, \quad \text{as } r \to 0$$

$$\sigma_{rz} = \mu u_{z,r} = 0 \quad \text{at } r = a$$

and compare the solution with the appropriate results given at the end of Section 5.2.

5-3 Deduce γ_{sc} for the Rayleigh region, $k_\beta a \ll 1$, for the case defined in Problem 5-1.

5-4 Propose a measure of scattering losses due to an elastic obstacle in a viscoelastic medium.

5-5 Is it possible to generalize the analysis given in Section 5.1 to the case of cylindrical incident waves?

5-6 The admittance of a rigid embedded sphere under the impact of harmonic P-waves is given by

$$H(i\omega) = 3\tilde{\rho}(-\chi^2 m^2 - 3i\chi m + 3)e^{-im}/\{\chi^2 m^4 \\ +i\chi[2\chi + 1 + \tilde{\rho}(\chi+2)]m^3 \\ -[2\chi^2 + 1 + \tilde{\rho}(\chi^2 + 9\chi + 2)]m^2 - 9i\tilde{\rho}(\chi+1)m + 9\tilde{\rho}\}$$

$$\tilde{\rho} = \rho_1/\rho_2, \quad m = k_{\alpha 1}a, \quad \chi = c_{\alpha 1}/c_{\beta 1} \quad .$$

Consider the sphere motion under the impact of random waves of white noise type.

5-7 Can local stresses in composites be found using the concept of an effective medium?

5-8 Find the axial shear modulus of a fibre-reinforced composite by the effective scatterer approach and by the composite cylinder assemblage.

5-9 One of the approximate equations for the effective bulk modulus of particular composites has the form

$$K_e = K_2 + (K_2 - K_1)\frac{3K_1 + 4\mu_1}{3K_2 + 4\mu_1}\phi \quad .$$

Improve the result by a differential scheme.

5-10 Show that for the direction cosines, α_{ij}, (see Section 5-9) the following result holds

$$\langle \alpha_{in}\alpha_{jn}\alpha_{kn}\alpha_{ln} \rangle = (\delta_{ij}\delta_{kl} + \delta_{ik}\delta_{jl} + \delta_{il}\delta_{jk})/15 \quad ,$$

where $\langle\ \rangle$ denotes angular averaging. Find $\langle c'_{1111} \rangle$.

5-11 Verify whether the Hashin-Strikman bounds for the bulk modulus coincide with the results of the composite spheres assemblage.

5-12 Apply the causal approach of independent scatterers to deduce the dynamic response of particulate composites.

5-13 Show in a qualitative way that (5.102) accounts for multiple scattering effects, unlike (5.101).

5-14 On letting d be a typical grain size the attenuation in polycrystals can be shown to follow the pattern:

$$\alpha(\omega) \sim 0(\omega^4), \quad \text{as} \quad \lambda/d \to \infty$$
$$\alpha(\omega) \sim 0(\omega^2), \quad \text{as} \quad \lambda/d \sim 0(1)$$
$$\alpha(\omega) \sim \text{constant}, \quad \text{as} \quad \lambda/d \to 0 \quad .$$

Verify whether the expression

$$\alpha(\omega) = A(\delta\omega - \sin\delta\omega)(\gamma\omega - \sin\delta(\omega)/(\gamma\delta\omega^2)$$

with constant A, γ, and δ complies with the above pattern. Find the wave speed.

5-15 Consider a transient response of the structure shown in Figure 5-15 (half-space) and outline an approximate method for its evaluation.

References and Additional Reading

Sections 5.1-5.4

Ying and Truell (1956) presented a classical solution for diffraction on a sphere. See also Gubernatis, Domany and Krumhansl (1977), Pao and Mow (1973). Visscher (1980) and Waterman (1976) proposed efficient methods for evaluating scattering cross-sections. See Beltzer (1980) for diffraction of random waves by a sphere and Mow (1965) for that of a transient wave.

Section 5.5

Christensen (1979), and the surveys of Hashin (1983), Watt, Davies and O'Connell (1976), and Willis (1981).

Section 5.6

The approach is due to Ament, for a modern treatment see Kuster and Toksoz (1974).

Section 5.7

The model was devised by Hashin, see Christensen (1979), Hashin (1983).

Section 5.8

McLaughlin (1977), Norris (1985). The idea is apparently due to Roscoe and to Bruggeman. The expression for the effective bulk modulus given in Problem 5-9 follows from the bounds derived by Miller.

Sections 5.9

See the references to Section 5.5.

Section 5.10

Musgrave (1970).

Section 5.11

See the references to Section 5.5. The monograph by Christensen (1979) and review by Willis (1981) contain derivations of the Hashin-Strikman bounds.

Section 5.12

Willis (1981). See Hodges (1982) for an introductory discussion of a phenomenon of localization.

Sections 5.13-5.15

The method was proposed by Beltzer (1983b) and then extended in Beltzer, Bert and Striz (1983), Beltzer and Brauner (1984, 1985a,b,1986), Beltzer (1986), Beltzer and Brauner (1988), and Beltzer (1988).

Section 5.16

Musgrave (1970), Hodges (1982).

Section 5.17

Christensen (1979), for an extensive analysis see Minagawa and Nemat-Nasser (1977), Sve (1971, 1972), Ziegler (1977).

Comments on Selected Problems

Chapter I

1-10 It is sufficient to consider the case with two symmetry planes only, say, x_2x_3 and x_1x_2. This implies that the x_1x_3 – plane is also a symmetry plane and the solid is orthotropic. Indeed, in addition to (1.37) we get

$$c_{16} = c_{26} = c_{36} = c_{45} = 0 \quad .$$

Thus, there are nine independent elastic constants.

1-12 Make use of the potential

$$G = U - E_i D_i - ST$$

and the following table:

$\alpha \setminus \beta$	σ_{ij}	D_i	S
$\epsilon_{k\ell}$	$c^{ET}_{ijk\ell}$	$e^{T}_{ik\ell}$	$\gamma^{E}_{k\ell}$
E_k	$-e^{T}_{kij}$	$\lambda^{\epsilon T}_{ik}$	p^{ϵ}_{k}
T	$-\gamma^{E}_{ij}$	p^{ϵ}_{i}	$C^{\epsilon E}/T$

1-19 Consider a differential element of a string, ds. The resultant force is $T\zeta ds$, where T is the tension and ζ the curvature. The force is normal to the wave profile. One the other hand, the acceleration is $c^2\zeta$, if c is the velocity of the particle. Is is supposed that a reference frame is chosen so that the wave profile is constant. Thus, this frame moves with the speed c, which becomes the phase velocity too. Equating the above forces we get

$$T\zeta ds = c^2 \zeta \rho ds$$

$$c^2 = T/\rho \quad ,$$

where ρ is the density per unit length.

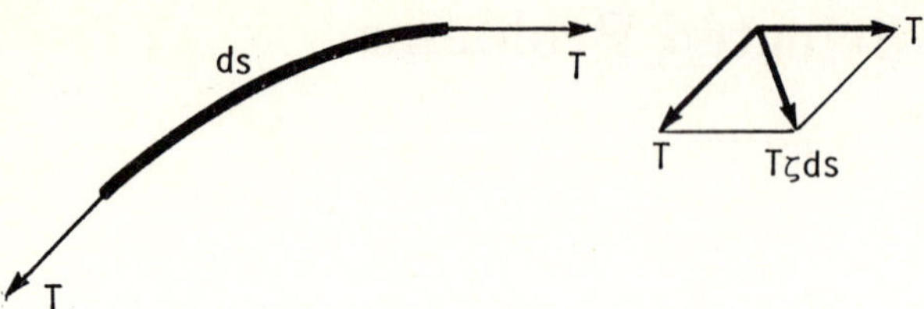

Fig. C-1 Forces acting on a string.

1-20 Consider the term $u(\omega, x)e^{-i\omega t}d\omega$ (see (1.94) and (1.95). Multiplying (1.76) by $e^{i\omega t}$ and integrating from c to d we get, first,

$$\int_c^d \ddot{u}e^{-i\omega t}dt = \frac{\partial u}{\partial t}e^{i\omega t}\Big|_c^d - i\omega u e^{i\omega t}\Big|_c^d - \omega^2 \int_c^d u e^{i\omega t}dt \qquad (A.1)$$

and second,

$$\int_c^d u_{,xx}\, e^{i\omega t}dt = u(\omega, x)_{,xx}\sqrt{2\pi} \quad .$$

For $u(t, x)$ periodic we put $T = d - c$, while $c \rightarrow -\infty$, $d \rightarrow \infty$, otherwise. Then only the last term in (A.1) survives, and we get

$$\int_c^d \ddot{u}e^{i\omega t}dt = -\omega^2 u(\omega, x)\sqrt{2\pi} \quad ,$$

which yields

$$u(\omega, x)_{,xx} + k^2 u(\omega, x) = 0 \quad .$$

1-22 In the new variable (1.76) takes the form

$$\frac{\partial^2 u}{\partial x'^2} - \frac{1}{c^2}\frac{\partial^2 u}{\partial t^2} + 2\frac{a}{c^2}\frac{\partial^2 u}{\partial t \partial x'} - \frac{a^2}{c^2}\frac{\partial^2 u}{\partial x'^2} = 0 \quad ,$$

which is not a wave equation. Thus, in a violation of the relativity principle, (1.76) assumes the existence of an "absolute" reference frame. This is due to an approximate nature of this equation.

Chapter II

2-8 From the general wave equation in spherical coordinates we get

$$c^2 \partial^2 (ru)/\partial r^2 - \partial^2 (ru)/\partial t^2 = 0 \quad .$$

2-15 Integrate (2.130) with respect to time from $-\infty$ to ∞.

2-17 Consider as an illustration the integral

$$\Pi(r) = \int_{-\infty}^{\infty} e^{ikr} dk/(k^2 - q^2) \quad ,$$

where we first put $\operatorname{Im} q \neq 0$. Here

$$k^2 = k_1^2 + k_2^2 + k_3^2 \quad .$$

After integration over the orientation of k we get

$$i \int_{-\infty}^{\infty} \frac{(e^{ikr} - e^{-ikr})kdk}{(k-q)(k+q)8\pi^2 r} \quad .$$

The residue theorem yields that $\Pi(r)$ is given by

$$\Pi(r) = e^{iqr}/(4\pi r) \qquad \text{for} \qquad \operatorname{Im} q > 0$$

$$\Pi(r) = e^{-iqr}/(4\pi r) \qquad \text{for} \qquad \operatorname{Im} q < 0 \quad ,$$

and is thus multivalued with a branch cut along the real axis. We may choose $\operatorname{Im} q \to 0^+$ on the basis of physical considerations concerning the viscoelastic attenuation in the medium.

2-22 Apply the operator

$$(2\pi i)^{-1} \int (...) d\omega/\omega$$

to (2.151) and (2.156) and use the relation

$$H_1^{(1)}(z) = -2 \int_1^{\infty} e^{izv} (v^2 - 1)^{-1/2} v dv/\pi \quad .$$

(Eshelby (1953)).

Chapter III

3-1 Apply a usual procedure for diagonalization of a symmetric matrix.

3-2 In this case

$$n_1 = n_2 = n_3 = 1/\sqrt{3}$$

and the answer for the P-mode is

$$V^2 = (c_{11} + 2c_{12} + 4c_{44})/(3\rho) \quad ,$$

while for the S-mode

$$V^2 = (c_{11} - c_{12} + c_{44})/(3\rho) \quad .$$

3-3 For the P-mode we get

$$V^2 = (c_{11} + c_{12} + 2c_{44})/(2\rho) \quad ,$$

while for the S-mode

$$V^2 = (c_{44}/\rho)^{1/2} \qquad \text{in the [001]-direction}$$

$$V^2 = (c_{11} - c_{12})/(2\rho) \quad \text{in the } [1\bar{1}0]\text{-direction} \quad .$$

Chapter IV

4-6 The displacements must vanish at the boundary.

4-9 Our analysis involves incident, reflected, and transmitted strain waves, which are supposed to be known from experiments. On neglecting the wave propagation effects within a short speciment we assume that the axial force, P, on each of its face is the same. This, in particular, yields the stress

$$\sigma = P/A = E\varepsilon_t \quad ,$$

where A is the rod cross-section and ε_t the transmitted strain. The displacements may be derived by integrating the strains.

4-12 The ideal fluid does not support any shear stress and would not therefore affect the Love waves.

4-17 Make use of the addition theorem of the cylindrical Hankel functions. (Beltzer and Parnes (1984)).

Chapter V

5-4 If the inclusion is elastic and the surrounding medium is viscoelastic, one should take into account that the intensity of incident field depends on the distance and is different for different points of a scatterer. Accordingly, the average power flux of scattered waves may be normalized by the spatially averaged intensity

$$\langle e^i \rangle_S = \frac{1}{ST} \int_0^T \int_S \int P_j^i m_j dt ds \quad .$$

Also, the integration in (5.19) must be carried out precisely over S, unlike the elastic case.

To complete the account for the energy perturbations we have still to evaluate the losses, which would occur at the site, if the obstacle were not present. These are given by the negative power flux over S evaluated for the incident field only

$$\langle I^a \rangle = -\frac{1}{T} \int_0^T \int_S \int P_j^i n_j dt ds \quad ,$$

which yields the absorption cross section, γ^a,

$$\gamma^a = \langle I^a \rangle / \langle e^i \rangle_S \quad .$$

The total effect is evaluated by the extinction cross section, γ^e,

$$\gamma^e = \gamma^s + \gamma^a = \langle I^s \rangle / \langle e^i \rangle_S + \langle I^a \rangle / \langle e^i \rangle_S \quad .$$

Since γ^a is negative, this relation suggests that the presence of an elastic obstacle in viscoelastic matrix may give rise to effects opposing the wave decay.

The concept of the extinction cross section enables us to modify (5.102) so as to account for the energy perturbations in a more accurate way. In fact, the current homogenized matrix is attenuative and dispersive. Since γ^e takes into account both scattering losses and absorption we get instead of (5.102)

$$\tilde{\alpha}(\phi + \Delta\phi) \simeq \tilde{\alpha}(\tilde{\phi}) + \Delta\tilde{\phi}\tilde{\gamma}^e/(2V) \quad .$$

(Beltzer and Brauner (1988)).

5-8 In each of the cylindrical regions the displacements are represented by

$$u_z = (Ar + B/r)\cos\theta$$

$$u_r = Cz\cos\theta$$

$$u_\theta = -Cz\sin\theta \quad ,$$

with $B = 0$ for the inner cylinder. The remaining coefficients are found from the boundary conditions at the interface and at the outer surface. The stresses are

$$\sigma_{rr} = \sigma_{r\theta} = \sigma_{\theta\theta} = \sigma_{zz} = 0$$

$$\sigma_{\theta\theta} = -\mu(A + C + B/r^2)\sin\theta$$

$$\sigma_{zr} = \mu(A + C - \beta/r^2)\cos\theta \quad .$$

The extension of the solution to the entire RVE is similar to that of the composite spheres assemblage.

5-10 It can be verified by summing with respect to i and then to j that

$$\alpha_{in}\alpha_{in}\alpha_{jn}\alpha_{jn} = 1 \quad .$$

Multiplying this by $\delta_{ij}\delta_{kl}$ we arrive at

$$\begin{aligned}\langle c'_{1111} \rangle &= \langle \alpha_{1i}\, \alpha_{1j}\, \alpha_{1k}\, \alpha_{1\ell} \rangle\, c_{ijk\ell} \\ &= (9c_{11} + 6c_{12} + 12c_{44})/15 \quad .\end{aligned}$$

5-13 On neglecting mutual interactions among inclusions we write the attenuation as

$$\alpha = (\gamma/2 + \gamma/2 + \ldots\gamma/2) = N\gamma/2 \quad .$$

This scheme is modified in the approach of the causal differential media by taking into account that scattering cross-sections of inclusions inserted at the i-step are influenced by inclusions inserted at the previous steps. Thus we arrive at (5.102).

5-14 Assume $\delta > \gamma$ and carry out limiting procedures. The phase velocity associated with the attenuation given by

$$\alpha(\omega) = A(\delta\omega) - \sin\delta\omega)(\gamma\omega - \sin\gamma(\omega)/\gamma\delta\omega^2)$$

with $\delta > \gamma$, may be obtained via the $K - K$ relation

$$c(\omega) = c_o\left[1 + \frac{2c_o\omega^2}{\pi}P\int_0^\infty \frac{\alpha(\omega')d\omega'}{\omega'^2(\omega'^2 - \omega^2)}\right]^{-1} \quad .$$

Here c_o follows from the static analysis,

$$c_o = \lim_{\omega\to 0} c(\omega) \quad .$$

Evaluation of the above integral in a complex plane with the poles $\omega' = 0$, $\omega' = \pm\omega$, yields

$$c_o/c(\omega) = 1 + Ac_o[-\cos(\delta\omega)/(\delta\omega^2) - \cos(\gamma\omega)/(\gamma\omega^2)$$
$$+ \sin(\gamma\omega)\cos(\delta\omega)/(\delta\gamma\omega^3) + 1/(\gamma\omega^2) - \gamma/2 + \gamma^2/(6\delta)] \quad ,$$

where $\delta > \gamma$, (Beltzer and Brauner (1985b)).

5-15 Consider the case of excitation of a laminated half-space by a step-function, Then for $x > 0$ the stress is

$$\sigma(x,t) = i\int_{-\infty}^{\infty} \omega^{-1}e^{i[k(i\omega)x - \omega t]}d\omega/(2\pi) \quad ,$$

where $k(i\omega) = k_r + ik_i$. The lowest mode of (5.120) may be expanded for low frequencies as

$$k_r(\omega) \simeq \omega/c_o + \xi\omega^2/c_o^2 \quad ,$$

with constant ξ. Adopting, say, the Kelvin-Voight model for viscolelastic losses, we similarly get

$$k_i(\omega) \simeq \beta\omega^2 \quad ,$$

with constant β. Substitution of these expansions for $k(i\omega)$ leads to an integral which may be evaluated by the convolution theorem

$$\int_{-\infty}^{\infty} g(\omega)f(\omega)e^{-i\omega t}d\omega = \int_{-\infty}^{\infty} f(\tau)f(t-\tau)d\tau \quad .$$

The solution holds for the far-field near the wave front, (Sve (1972)).

References

ACHENBACH, J.D. (1973) Wave Propagation in Elastic Solids, North-Holland, Amsterdam.

AULD, B.A. (1973) Acoustic Waves in Solids, Vols. 1 and 2, Wiley-Interscience, New York.

BEDFORD, A. (1985) Hamilton's Principle in Continuum Mechanics, Pitman, Boston.

BELTZER, A.I. (1980) Random Response of a Rigid Sphere Embedded in a Viscoelastic Medium and Related Problems, J. Appl. Mech., Vol. 47, No. 3, 499-503.

BELTZER, A.I. (1981) Random Rayleigh Waves in Viscoelastic Media, J. Acoust. Soc. Am., Vol. 70, No. 5, 1357-1361.

BELTZER, A.I. (1983a) Radiation from a Dislocation Oscillating in a Circular Cylinder, Int. J. Engng. Sci., Vol. 21, No. 2, 165-170.

BELTZER, A.I. (1983b) Kramers-Kronig Relationships and Wave Propagation in Composites, J. Acoust. Soc. Am., Vol. 73, No. 1, 355-356.

BELTZER, A.I. (1986) Causality and the Keller Approximation for Acoustic Waves, J. Appl. Phys., Vol. 59, No. 5, 1986, 1456-1457.

BELTZER, A.I. (1988) Dispersion of Seismic Waves By a Causal Approach, Pure and Appl. Geophys. (to appear).

BELTZER, A.I., C. BERT AND A. STRIZ (1983) On Wave Propagation in Random Particulate Composites, Int. J. Solids and Structs., Vol. 19, No. 9, 785-791.

BELTZER, A.I. AND N. BRAUNER (1984) Waves of Arbitrary Frequency in Random Fibrous Composites, J. Acoust. Soc. Am. Vol. 76, No. 3, 962-963.

BELTZER, A.I. AND N. BRAUNER (1985a) SH-Waves of an Arbitrary Frequency in Random Fibrous Composites via the K-K Relations, J. Mech. Phys. Solids, Vol. 33, No. 5, 471-487.

BELTZER, A.I. AND N. BRAUNER (1985b) Elastic Waves Propagation in Polycrystalline Media: A Causal Response, Appl. Phys. Letters, Vol. 47, No. 10, 1054-1055.

BELTZER, A.I. AND N. BRAUNER (1986) Acoustic Waves in Random Discrete Media via a Differential Scheme, J. Appl. Phys., Vol. 60, No. 2, 538-540.

BELTZER, A.I. AND N. BRAUNER (1987) The Dynamic Response of Random Composites by a Causal Differential Method, Mechanics of Materials, Vol. 6, No. 4, 337-395.

BELTZER, A.I. AND R. PARNES (1984) Resonance Radiation from Imperfections, Part I, Mechanics of Materials, Vol. 3, 199-210.

BEN-MENACHEM, A. AND D. SINGH, (1981) Seismic Waves and Sources, Springer- Verlag, New York.

BERT, C.W. (1973) Material Damping: An Introductory Review of Mathematical Models, Measures and Experimental Techniques, J. Sound Vibr. Vol. 29, 129-153.

BIDERMAN, V.L. (1972) Applied Theory of Mechanical Vibrations (in Russian: Prikladnaye Teoriey Mechanicheskih Kolebanei) Vuschaey Shkola, Moscow.

CHADWICK, P. AND E.A. TROWBRIDGE (1967) Oscillations of a Rigid Sphere Embedded in an Infinite Elastic Solid, Proc. Camb. Phil. Soc. (Math & Phys. Sciences), Vol. 63, 1189-1227.

CHRISTENSEN, R.M. (1971) Theory of Viscoelasticity, Academic Press, New York.

CHRISTENSEN, R.M. (1979) Mechanics of Composite Materials, John Wiley and Sons, New York.

DIEULESAINT, E. AND D. ROYER, (1980) Elastic Waves in Solids, John Wiley & Sons, Toronto.

EASON, G., J. FULTON AND I.N. SNEDDON (1955-1956) The Generation of Waves in an Infinite Elastic Solid by Variable Body Forces, Phil. Trans. of the Royal Soc., London, Series A248, 575-607.

ESHELBY, J.D. (1949) Dislocations as a Cause of Mechanical Damping in Metals, Proc. Royal Soc. Series A, Vol. 197, 396-416.

ESHELBY, J.D. (1953) The Equation of Motion of a Dislocation, Phys. Review, Vol. 90, No. 2, 248-255.

ERINGEN, A.C. and E.S. SUHUBI (1974-1975) Elastodynamics, Vol. 1 and 2, Academic Press, New York.

FEDOROV, L. (1973) Theory of Elastic Waves in Crystals, Plenum Press, New York.

FRYBA, L. (1973) Vibrations of Solids and Structures Under Moving Loads, Martinus Nijhoff, Hague.

GRAFF, K.F., (1975) Wave Motion in Elastic Solids, Ohio State Univ. Press.

GREEN, A.E. and W. ZERNA (1968) Theoretical Elasticity, Oxford Univ. Press, New York.

GUBERNATIS, J.E., E. DOMANY AND J.A. KRUMHANSL (1977) Formal Aspects of the Theory of the Scattering of Ultrasound by Elastic Materials, J. Appl. Phys., Vol.48, 2804-2811.

HASHIN, Z. (1983) Analysis of Composite Materials, J. Appl. Mech., Vol. 50, No. 3, 481-505.

HUTCHINSON, J.R. (1984) Vibrations of Thick Free Circular Plates, Exact Versus Approximate Solutions, J. Appl. Mech. Vol. 51, 581-585.

HODGES, C.H. (1982) Confinement of Vibration by Structural Irregularity, J. Sound Vib., Vol. 82, 411-424.

KOLSKY, H. (1963) Stress Waves in Solids, Dover Publ., New York.

KOSEVICH, A.M. (1979) Crystal Dislocations and the Theory of Elasticity, in: Dislocations in Solids, edited by F.R. N. Nabarro, Vol. 1, North-Holland, Amsterdam.

KUBO, R. AND T. NAGAMIYA (1969) Solid State Physics, McGraw-Hill, New York.

KUSTER, G.T. AND M.N. TOKSOZ (1974) Velocity and Attenuation of Seismic Waves in Two-Phase Media, Geophysics, Vol. 39, No. 5, 587-618.

MARIANI, E.A. (1985) Terms and Definitions for SAW Devices, Sonics and Ultrasonics, IEEE Trans., Vol. SU-32, No. 4, 476-480.

McLAUGHLIN, R. (1977) A Study of the Differential Scheme for Composite Materials, Int. J. Engng. Sci., Vol. 15, 237-244.

MEEKER, T.R., AND A.H. MEITZLER (1964) Guided Wave Propagation in Elongated Cylinders and Plates, in Physical Acoustics, edited by W.P. Mason, Vol. 1, Part A, Academic Press, New York, 111-167.

MIKLOWITZ, J. (1978) The Theory of Elastic Waves and Waveguides, North-Holland Publ. Co., Amsterdam.

MINAGAWA, S. AND S. NEMAT-NASSER, (1977) On Harmonic Waves in Layered Composites, J. Appl. Mech., Vol. 44, 689-695.

MINDLIN, R.D. (1951) Influence of Rotary Inertia and Shear on Flexural Motions of Isotropic, Elastic Plates, J. Appl. Mech. Vol. 18, 31-38.

MORSE, P.M. AND K.U. INGARD (1968) Theoretical Acoustics, McGraw-Hill, New York.

MOW, C.C. (1965) Transient Response of a Rigid Sperical Inclusion in an Elastic Medium, J. App. Mech., Vol. 32, No. 3, 637-642.

MURA, T. (1982) Micromechanics of Defects in Solids, Martinus Nijhoff, Hague.

MUSGRAVE, M.J.P. (1970) Crystal Acoustics, Holden Day, San-Francisco.

NORRIS, A.N. (1985) A Differential Scheme for the Effective Moduli of Composites, Mechanics of Materials, Vol. 4, 1-16.

NOWACKI, W.K. (1978) Stress Waves in Non-Elastic Solids, Pergamon Press, Oxford.

NUSSENZVEIG, H.M. (1972) Causality and Dispersion Relations, Academic Press, New York.

PAO, Y-H, AND C-C. MOW, (1973) Diffraction of Elastic Waves and Dynamic Stress Concentrations, Grane Russak, New York.

PARNES, R. AND BELTZER, A.I. (1984) Resonance Radiation from Imperfections, Part II, Mechanics of Materials, Vol. 3, 211-221.

PARNES, R. AND BELTZER, A.I. (1986) Effect of Viscoelastic Losses on Resonance Radiation from Imperfections, Ultrasonics, Vol. 24, 189-196.

PIPKIN, A.C. (1972) Lectures on Viscoelasticity Theory, Springer-Verlag, New York.

PURSEY, H. (1957) The Launching and Propagation of Elastic Waves in Plates, Quart. J. Mech. and Applied Math., Vol. 10, 45-62.

REISMANN, H. (1967) On the Forced Motion of Elastic Solids, Appl. Sci. Res., Vol. 18, 156-165.

REISMANN, H. (1968) Forced Motion of Elastic Plates, J. Appl. Mech., Vol. 35, 510-515.

SACHSE, W. AND N.N. HSU (1979) Ultrasonic Transducers for Materials Testing and Their Characterization, in: Physical Acoustics, edited by W.P. Mason and R.N. Thurston, Vol. 14, Academic Press, New York. 277-406.

SETH, B.R. (1966) Measure-Concept in Mechanics, Int. J. Non-Linear Mechanics, Vol. No. 1, 35-40.

SOKOLNIKOFF, I.S. (1956) Mathematical Theory of Elasticity, McGraw-Hill, New York.

SVE, C. (1971) Time-Harmonic Waves Travelling Obliquely in a Periodically Laminated Medium, J. Appl. Mech., Vol. 38, 477-482.

SVE, C. (1972) Stress Wave Attenuation in Composite Materials, J. Appl. Mech., Vol. 39, 1151-1153.

TUCKER, J.W. AND V.W. RAMPTON (1972) Microwave Ultrasonics in Solid State Physics, North-Holland, Amsterdam.

VAN VLACK, L.H. (1980) Elements of Materials Science and Engineering, Addison- Wesley, Reading.

VIKTOROV, I.A. (1967) Rayleigh and Lamb Waves, Plenum Press, New York.

VISSCHER, W.M. (1980) A New Way to Calculate Scattering of Acoustic and Elastic Waves, I and II, J. Appl. Phys. Vol. 51, No. 2, 825-845.

WATERMAN, P.C. (1976) Matrix Theory of Elastic Wave Scattering, J. Acoust. Soc. Am., Vol. 60, No. 3, 567-580.

WATT, J.P., G.F. DAVIES AND R.J. O'CONNELL (1976) The Elastic Properties of Composite Materials, Reviews of Geophysics and Space Physics, Vol. 14, No. 4, 541-563.

WEERTMAN, J. AND J.R. WEERTMAN (1980) Moving Dislocations, in: Dislocations in Solids, edited by F.R.N. Nabarro, Vol. 3, North-Holland, Amsterdam.

WILLIS, J.R. (1981) Variational and Related Methods for the Overall Properties of Composites, in: Advances in Appl. Mech., Vol. 21, Yih, C.S. ed. Academic Press, New York, 1-78.

YING, C.F. AND R. TRUELL (1956) Scattering of a Plane Longitudinal Wave by a Spherical Obstacle in an Isotropically Elastic Solid, J. Appl. Phys., Vol. 27, No. 9, 1086-1097.

ZIEGLER, F. (1977) Wave Propagation in Periodic and Disordered Layered Composite Elastic Materials, Int. J. Solids and Structs., Vol. 13, 293-305.

Author Index

Subject Index